Composite Steel and
Concrete Construction

Composite Steel and Concrete Construction

P. R. KNOWLES

Lecturer in Civil Engineering,
University of Surrey

A HALSTED PRESS BOOK

JOHN WILEY & SONS

New York - Toronto

Published in the U.S.A. and Canada by Halsted Press,
a Division of John Wiley & Sons, Inc., New York

Library of Congress Cataloging in Publication Data

Knowles, Peter Reginald.
   Composite steel and concrete construction.

" A Halsted Press book."
   1. Composite construction. I. Title.
TA664.K57 1973   ·   624′.1821   73-11317
ISBN 0-470-49580-4

Printed in Hungary

Preface

Concrete and steel are used structurally in such large quantities that methods of design, fabrication or construction that can show economies in final cost are of great utility. The object of this book is to present one combined application of the two materials in which they are intimately linked to form composite structural elements. The emphasis of the book is towards the engineering applications of composite construction through an explanation, in general terms, of the principles of the subject, followed by sections on the uses of composite construction in buildings and bridges.

The literature devoted to the technologies of structural steel and concrete is so vast that it is possible in this book to present from it only those facts which are essential to an understanding of composite action. For fuller information the reader is referred to the many excellent texts on one or the other material.

The bulk of this book has been compiled from many and various sources. To these sources I gratefully express my indebtedness, hoping that acknowledgment of this debt has been adequately made in the reference lists at the end of each Chapter.

Mr Derrick Beckett and Mr Leslie Leech, who generously read the manuscript, have made many helpful suggestions. In thanking them and Mrs D. Claxton who typed the manuscript I reserve to myself the responsibility for any errors that this book may contain.

P.R.K.

Contents

CONTENTS

CHAPTER 1

Composite Construction

1.1 Introduction

The term 'composite construction', used without qualification, can refer to structural systems in which there is interaction between such diverse materials as steel and concrete in reinforced concrete, steel and timber in a flitched beam, glass fibres in a resin matrix, or brick infill panels in a steel frame. In this book the term is used solely to refer to interaction between concrete and structural steel in such combinations as a steel beam or truss interacting compositely with a reinforced concrete slab. In bridges and buildings concrete deck and floor slabs are the most common load distributing medium, transferring load superimposed on them to a supporting structure of steel beams. Composite structural systems are therefore important, accounting for a considerable proportion of the annual construction output. Attempts to improve the efficiency of these systems will have clear economic advantages and indeed these can be demonstrated.

While the most commonly used form of composite system is that employing solid web steel beams supporting concrete slabs, many other types have been used (see Figure 1.1). The supporting beam may be an open web joist, a castellated beam, a truss or a box beam. The concrete slab may partially or wholly enclose the beam or, where a steel box beam is used, form the top flange of the box. Also included in the general classification of composite construction in steel and concrete is the composite column, which is typically a steel tube filled with concrete. The only requirement is that composite action must be ensured by means of suitable shear connection unless the natural bond between steel and concrete can be relied on. Conventional reinforced or prestressed concrete, although composite materials, fall outside the scope of composite construction as described here because the steel reinforcement is not structurally self-supporting. The steel

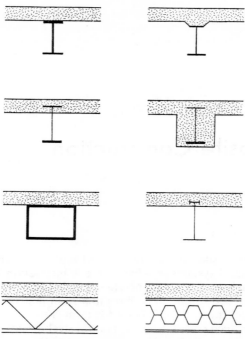

Figure 1.1 Types of composite beam

beams in composite construction, on the other hand, are capable of supporting their own dead weight and that of the associated concrete, thus leading to considerable simplification of the construction process.

1.2   Early development

The history of composite construction is intimately connected with that of reinforced concrete, of which it is indeed but a special case. Although weak concrete was made by the Romans with naturally occurring cements, they did not have cause to use it in situations where it might be in tension. The discovery of a strong hydraulic cement by Aspdin in 1824 was the necessary precursor of the better quality cement needed for effective reinforced concrete. A patent for reinforced concrete in building was taken out in 1854 and isolated examples of the use of the material in building structures occurred in England after 1860. However, attempts to make use of the *structural* properties of reinforced concrete, in England at any rate, met with little success, concrete being regarded mainly as a convenient material for fireproofing structural iron or steelwork.[1]

Disastrous fires in timber buildings were common in the Middle Ages, but the effects were much less acute than in the later period at the beginning of the Industrial Revolution when towns were expanding

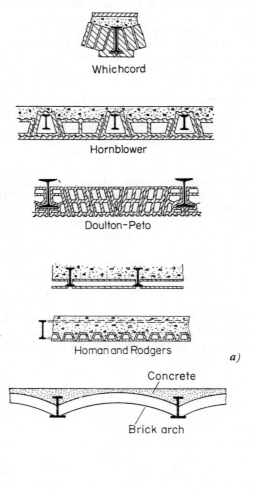

Whichcord

Hornblower

Doulton-Peto

Homan and Rodgers

a)

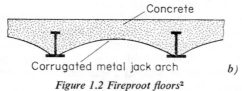

Concrete

Brick arch

Concrete

Corrugated metal jack arch

b)

Figure 1.2 Fireproof floors[2]

and factories with high fire risks were being built. When the use of iron framing became a possibility about 1790, with the production of the first iron beams and columns, it must have seemed to engineers and architects that the fireproof building had become a reality. However, the effect of fire on iron frames was soon discovered to be severe; incombustible they might be but fire reduced their strength to practically nothing. Fire protection of the iron beams and columns was a necessity and from 1800 onwards numerous fireproofed floor systems were proposed, some of which are shown in Figure 1.2[2]. Of immediate interest are those systems employing concrete (Figure 1.2(b).). Generally the concrete weight was carried by brick jack arches, iron sheeting or other means into the floor beams so that it did not act structurally in spanning between the beams but simply as fireproofing to the metal work. The weight of the floor imposed a severe loading on its supporting beams without any compensating structural action.

It was not a long step from the concrete floor supported on jack arches to the filler joist floor consisting of iron or steel beams encased in a slab of concrete. While the concrete in fact provided the floor (or bridge deck for the same system was used in bridges) its main function was to protect the beam from fire and corrosion. Design methods in use at the time of development of the fireproof floor realistically treated brick arches and floor structure as a dead load carried by beams; it was inevitable that the filler joist floor should be unrealistically treated in the same way. By 1900 reinforced concrete was in structural use in England but the design formulae for it did not cater for reinforcement of such large relative size as a filler joist. There was indeed a conviction in building circles that filler joist floors were 'half as strong again' as the steel beams alone but there were no reliable test results to prove the statement. In terms of structural efficiency a very high price was being paid for fireproof construction.

Just before the war of 1914–18 the structural steel firm of Redpath Brown and Company decided to institute a series of tests at the National Physical Laboratory which would provide authoritative data on the strength of filler joist floors, although because of the war the tests were not completed until 1923. The tests did show the increased strength which accompanies composite action and in 1925 Scott,[3] who had originally suggested the experimental investigation, presented an analysis of the test results, deriving also equations which could be used to design filler joist floors. In a later paper Caughey and Scott[4] extended the latter's earlier design theory to deal with the problem of a concrete slab resting on the top flange of a steel beam. This was a particularly significant development, especially as they pointed out the need for some form of mechanical shear connection between beam and slab suggesting that this could be provided 'by some simple device, such as riveted or welded angle clips or projecting bolt ends spaced at intervals from the ends of the beam as required to pick up

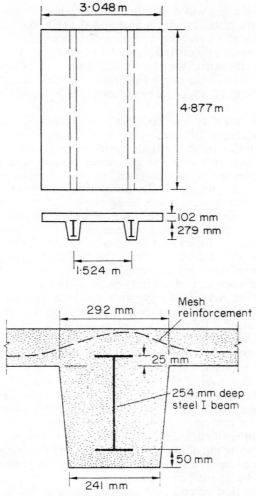

Figure 1.3 Canadian beam test specimen, 1922

the shear'. Deflection and shear were treated, design charts were given and a distinction was made between the loading case when the slab was propped until it had hardened and the unpropped case.

By the type of coincidence that often seems to happen in scientific research a series of tests on a composite beam and slab were carried out in Canada in 1922[5] under the direction of the Dominion Bridge Company. The test slab is shown in Figure 1.3. In some ways the tests were more advanced than those at the National Physical Laboratory since loading was distributed in such a way as to cause negative bending in the slab over the beams and the slab was reinforced with bars. Additionally a two day vibration test, in which a concrete mixer was run continuously on the slab, was carried out. The investigation report gave formulae for the elastic design of composite beams, making the interesting point that the commonly used value of the modular ratio of fifteen could be reduced if high strength concrete were used.

By 1930 it could be said that the initial test work establishing the composite strength of steel beam and concrete slab systems was complete. Elastic design methods based on conventional beam theory had been developed and the composite system was beginning to attract the attention of designers and inventors. Research workers, too, were taking an interest in the unresolved problems of composite action, much fundamental work being carried out at research establishments in a number of countries. A comprehensive review of international research work covering the period from the early experiments until 1960 was published by Viest[6], and a similar review by Johnson[7] covered the period 1960 to 1969. The deliberate use of composite action between steel and concrete must have been made soon after the early test work carried out in England and Canada in 1922, though the application was probably with concrete encasement of steel sections which would have been less noteworthy than externally 'augmented steel'. (This term was proposed by Scott[3] as a convenient way of distinguishing composite construction from reinforced concrete; it did not, however, become widely used.) In an industry not particularly addicted to change the adoption of a new method is likely to be slow even if the method can be shown to be economical. However, bridges and buildings in composite construction were erected in many countries in the period 1922–39. In Australia a paper by Knight[8] on composite slab and girder bridges, presented as early as 1934, dealt with such detailed points as the design of shear connectors, the propping of main girders, the effect of varying the modular ratio on the composite section properties and the prestressing of the steel girders by upward cambering.

The publication of codes of practice tends to lag behind current engineering thought. It is therefore not surprising that the first national code (the American Association of State Highway Officials (AASHO) specification for composite highway bridges) was not published until 1944. With its publication, official recognition and an approved design method were available to designers and so a yearly increasing number of composite highway bridges was built in the USA. The soundness of the basic design principles was demonstrated but a large volume of concurrent research showed that there was a need for attention to certain details. For this reason a new AASHO specification was published in 1957 and again in 1961.

The other major country to develop an interest in composite construction to the extent of publishing a Code of Practice (DIN 1078) was Germany. Here the pressures of steel shortage immediately following 1945 forced engineers to adopt the most economical design methods available in order to be able to cope with the very large amount of reconstruction of bridges and buildings destroyed in the 1939–45 war.

With two major codes of practice available (the American and German) other countries were able to draw up their own, basing them generally on the provision of one or other of the major codes. The AASHO code was in many ways much simpler than the German; possibly because steel was much more easily obtained in the USA the Americans did not attempt the structural complexity or detailed analysis found essential by the Germans. They preferred to avoid the difficulties of using continuous beams or prestressed decks which in German practice were necessary in order to show maximum economy in material.

1.3 Economy and other advantages

Where composite action can be achieved, certain structural advantages immediately appear. In comparison with the non-composite case the advantages may be summarised as:

(a) A reduction in steel area required to support a given load
(b) On a static ultimate load basis, an increase in the overload capacity over that of a non-composite beam (although fatigue effects may reduce this)
(c) For a given load, a reduction in construction depth with consequent saving in embankment costs for bridges or storey height in buildings

The first two factors will inevitably lead to a reduction in the steel weight required to support a given load, although the reduction may be offset to some extent by the need to provide shear connection for composite action. It does appear, then, that composite structures will show economy over their non-composite counterparts and this is generally borne out in practice.

Comparative design studies carried out in the USA[9] for simple bridge spans from 9·1 m to 27·4 m, with beam spacings of 1·52 m to 2·13 m, gave results, in terms of steel weight, summarised in *Table* 1.1. (The steel weights include an allowance in the case of composite beams for the weight of adequate shear connection.) Reduction in steel weight does not, of course, guarantee reduction in overall cost. However, available cost comparisons do show savings of the order of ten to twenty per cent. It will be seen from the table that 'propped'

Table 1.1 RELATIVE WEIGHTS OF SIMPLE BRIDGE BEAMS

Type of beam	Relative weight (%)
Non-composite rolled beam	100
Composite symmetrical rolled beam (a) without flange plates	
(i) unpropped	92
(ii) propped	77
(b) with flange plates on bottom flange	
(i) unpropped	76
(ii) propped	64
Composite welded plate girder*	
(a) unpropped	69
(b) propped	40–60

* Unsymmetrical, larger tension flange

construction generally shows the greatest reduction in steel weight. This method of construction involves supporting the steelwork from below with suitable props so that the whole of the dead load of steel and concrete is taken by the props which remain in position until the concrete has cured. By this means full composite action is available for dead load, in contrast with 'unpropped' construction in which the dead weight of steel beam and concrete slab is taken by the steel beam alone. For bridges, propping may be difficult or impossible, and in buildings props may be a hindrance to operations such as screeding and finishing which can proceed very soon after the floor slab is cast if the floor area is free of obstruction.

The overall economics of more complex bridge systems employing continuous girders for which special arrangements are required to eliminate tension in the concrete deck under negative bending are less clear, although the limited objective of reduction in steel weight will certainly be achieved. However, the figures given in *Table* 1.2

Table 1.2 COMPARISON OF CONTINUOUS BEAM SOLUTIONS FOR THREE-SPAN BRIDGE OF TOTAL LENGTH 96 m

Type of beam	Weight (%)	Cost (%)
Non-composite	100	100
Composite	87	92

for a continuous composite bridge in Japan show economy in both steel and overall cost.[10] The costs were for a continuous bridge of three spans of 32 m each.

Table 1.3 COMPARATIVE WEIGHT AND COST–SINGLE-BAY FIVE-STOREY FRAMES

Type of frame	Weight (%)	Cost (%)
1 Elastic non-composite	100	100
2 Elastic composite	84·5	92·5
3 Plastic non-composite	89	95·5
4 Plastic non-composite	87	93
5 Plastic composite	70	87
6 Plastic composite	71	85

For the economics of composite construction in building, a series of design studies are available comparing composite and non-composite frames designed elastically and plastically.[11] For a single-bay five-storey frame the relative weights and costs are shown in *Table* 1.3; for a larger frame see *Table* 1.4.

Table 1.4 COMPARATIVE WEIGHT AND COST–THREE-BAY SIX-STOREY FRAMES

Type of frame	Weight (%)	Cost (%)
1 Elastic non-composite	100	100
2 Elastic composite	86	91
3 Plastic non-composite	95	102
4 Plastic composite	66	90

The greatest economy in steel weight arising from composite action is achieved by methods of erection employing a form of prestressing of the steel beam. An idea of the reduction in the weight of steel beams in a typical building frame can be gained from *Table* 1.5. It is claimed

Table 1.5 RELATIVE WEIGHTS OF STEEL BEAMS IN A BUILDING FRAME

Type of beam	Relative weight (%)
Non-composite	100
Composite, propped	73
Composite, prestressed	55

that the prestressing method employed (described in detail in Chapter 6) adds very little to the constructional costs of the building.[12]

2

Over the past fifty years the very large amount of theoretical, experimental and constructional work carried out on composite structural systems has amply proved the efficiency and economy of the method. It can be said that where concrete slabs and steel beams are to be used in conjunction there is generally an advantage to be gained in making the two into a composite whole.

REFERENCES CHAPTER 1

1. BOWLEY, M., *The British building industry*, Cambridge University Press, p. 15, (1966)
2. HAMILTON, S. B., ' A short history of the structural fire protection of buildings particularly in England', National Building Studies Special Rep. No. 27, HMSO, London (1958)
3. SCOTT, W. B., 'The strength of steel joists embedded in concrete', *Struct. Engr*, No. 26, 201–219 and 228, (1925)
4. CAUGHEY, R. A. and SCOTT, W. B. 'A practical method for the design of I-beams haunched in concrete', *Struct. Engr*, **7**, No. 8, 275–93 (1929)
5. GILLESPIE, P. MACKAY, H. M. and LELUAU, C., Report on the strength of I-beams haunced in concrete', *Engng J.*, **6**, No. 8, pp. 365–69, Montreal (1923).
6. VIEST, I. M., Review of research on composite steel-concrete beams,' *Proc. Am. Soc. civ. Engrs., Struct. Div.* 1–21, (June 1960)
7. JOHNSON, R. P. 'Research on steel-concrete composite beams', *J. Struct. Div. Am. Soc. civ. Engrs*, **96**, No. ST3, 445–459 (Mar. 1970).
8. KNIGHT, A. W., 'The design and construction of composite slab and girder bridges', *J. Instn Engrs Aust.*, **6**, No. 1 (1934).
9. SIESS, C, P., 'Composite construction for I-beam bridges', *Trans. Am. Soc. civ. Engrs*, **114**, 1023–1045 (1949)
10. IWAMOTO, K., 'On the continuous composite girder', *Highway Res. Board Bull.* 339. (Bridge deck design and loading studies p. 81.) National Academy of Sciences National Research Council, Washington D.C. (1962)
11. JOHNSON, R. P., FINLINSON, J. C. H., and HEYMAN, J., 'A plastic composite design, *Proc. Instn civ. Engrs*, **32**, 198–209 (Oct. 1965)
12. WILENKO, L. K., The application to large buildings of structural steelwork prestressed during erection' *Proc. Conf. on steel in architecture*, British Constructional Steelwork Association, London (1969)

CHAPTER 2

Fundamentals of Composite Action

2.1 Introduction

The evolution of satisfactory design methods for composite beams has been a slow process, requiring much theoretical and experimental work in order to provide economic and, at the same time, safe design criteria. The purpose of this Chapter is to describe in some detail the more important fundamentals which have to be taken into account in the design of composite structures.

Historically, the first analysis of a composite section was based on the conventional assumptions of the elastic theory which limit the stresses in the component materials to a certain proportion of their 'failure' stresses (yield in the case of steel, crushing in the case of concrete). The assumptions inherent in the elastic method are similar to those for ordinary reinforced concrete.[1] In recent years the concepts of the ultimate load design philosophy have been applied to composite action, and a body of experimental evidence has shown it to be a safe, economical basis on which to proportion composite sections. Although at the present time ultimate load design methods are directly applicable only to buildings and not to bridges there seems no reason to doubt that in time the restriction will disappear.

Before dealing in detail with the two design approaches (elastic and ultimate load) certain basic points require consideration.

2.2 Properties of materials

A clear understanding of the way in which the component materials, steel, concrete and shear connection, react to applied load is an essential preliminary to full analysis of the composite section. Of primary importance are the stress–strain relationships, which must of necessity be the product of carefully controlled experiment. These experimental results are not generally suited to direct application and so simplifications and idealisations are adopted in practice. The use of computers has made it possible to reduce the amount of idealisation required with the result that computer 'experiments' can now be performed using material stress–strain relationships of considerable complexity.

2.2.1 STEEL

The steel beam component of a composite section may be fabricated entirely from one grade of steel, or from a combination of two grades (a *hybrid* beam). Additionally the composite section may contain steel reinforcement in the slab and prestressing steel in the slab or on the steel beam.

Steel beam material has the typical experimental stress–strain curve shown in Figure 2.1, exhibiting a yield point at A, a region of plastic extension AB and strain hardening from B to E. The ductility of structural steel allows considerable strain to occur before failure.

The experimental curve is awkward to describe mathematically, particularly in the plastic range, and so one of the idealised curves of

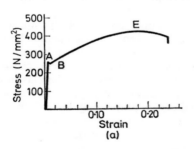

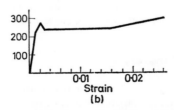

Figure 2.1 Stress–strain curve for mild steel beam material

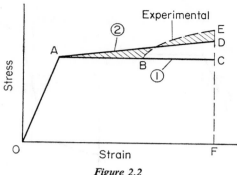

Figure 2.2

Figure 2.2 is usually adopted. Curve 1, OABC, makes no allowance for strain hardening unlike curve 2, the slope of which is adjusted to give equal intercepted areas between it and the experimental curve over the expected strain range OF. In practice curve 1 describes the behaviour of the steel adequately; neglect of strain hardening generally produces a higher load factor in the structure than that calculated by analysis.

Typical mechanical properties of two types of steel in common use in Great Britain are given in *Table* 2.1. Elongation, a measure of

Table 2.1 PROPERTIES OF STEEL TO BS 4360

Grade	Yield stress (N/mm²)	Ultimate tensile stress (N/mm²)	Elongation (%)
43	250	433/510	20
50	350	494/618	18

ductility, is the increase in length at failure as a proportion of the original length. A value of the elastic modulus E_s of $2 \cdot 1 \times 10^5$ N/mm² commonly adopted.

Ordinary steel reinforcement may be taken to have the same value of E_s of $2 \cdot 1 \times 10^5$ N/mm² but the stress–strain curves are of two distinct types; those for as-rolled material (Figure 2.3(a)) show a clear yield point, while cold-worked steel does not do so. For the latter the basic mechanical property is the 0·2 % proof stress (Figure 2.3(b)).

Prestressing steel is supplied as wire, bar, strand or cable, the stress–strain curves being similar to those for cold-worked steel reinforcement The value of E depends on the type of prestressing steel as shown in *Table* 2.2.

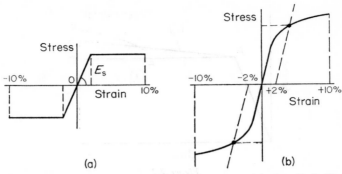

Figure 2.3 Stress–strain curves for steel reinforcement. (a) As rolled; (b) Cold worked[2]

Table 2.2 MODULUS OF ELASTICITY OF PRESTRESSING STEEL

Prestressing tendon	E_s (N/mm²)
Single section (wire or bar)	$2{\cdot}0 \times 10^5$
Strand	$1{\cdot}95 \times 10^5$
Cable	$1{\cdot}85 \times 10^5$

2.2.2 CONCRETE

Complexities in the analysis of composite sections are to a large extent the result of the complex nature of the material properties of concrete and in particular the time-dependence of its response to loading. The strain at any instant in concrete is composed of a mixture of elastic and plastic effects, dependent not only on the previous loading history but also on (to mention but some of the many possible causes of strain in the material) such diverse factors as the ambient conditions, the relative thickness of the concrete and its composition. Time-dependent effects are considered in sections 2.7–2.9; in this section only loadings of brief duration are examined.

Typical short term stress–strain curves for three different concretes having compressive strengths of 20, 40 and 60 N/mm² are shown in Figure 2.4. From the curves the value of the instantaneous modulus of concrete may be found by measuring the slope of the tangent to the curve at the origin. It may also be calculated from the relationship

$$E_c = 6{\cdot}0 \times 10^3 \, (f_{cu})^{1/2}$$

where f_{cu} is the cube strength of the concrete in N/mm².

For elastic design, concrete stress may be assumed proportional to strain, the appropriate modulus of elasticity being used. It will be seen

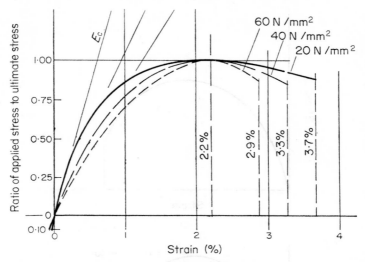

Figure 2.4 Short term stress–strain curves for concretes of different cube strengths[2]

from Figure 2.4 that, provided that the stresses do not exceed about 0·4 of the compressive strength of the concrete, the stress–strain curve may reasonably be assumed to be a straight line. (Working load stresses are generally of the order of 0·3 of the concrete strength). At ultimate load the assumption of a straight line relationship is clearly not justifiable, but the use of the curve of Figure 2.4 is cumbersome and so a simplified stress–strain curve is adopted; this will simulate as nearly as possible the static effect (position and magnitude of the resultant force in concrete) of the stress across the concrete. Two curves are shown in Figure 2.5; curve (a) is a combination of a parabola and rectangle, curve (b) is a trapezoidal function.

2.2.3 SHEAR CONNECTION

The important property of shear connection is its load–slip characteristic, which varies widely with the type of shear connector employed. In practice the slip is often ignored, the connection being treated as if it were infinitely rigid. Where required, the load–slip characteristic (see Figure 2.6) may be idealised in the form[3]

$$N = \alpha(1 - e^{-\beta s})$$

where N and s are the load and slip respectively, α and β are experimentally determined constants and e is the exponential function.

16

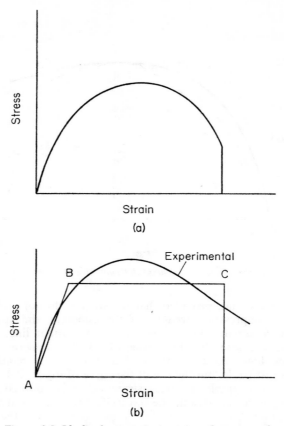

Figure 2.5 Idealised stress–strain curves for concrete[2]

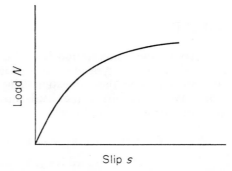

Figure 2.6 Load–slip curve for shear connector

2.3 Effective width

A composite beam and slab system is essentially a ribbed plate or a series of interconnected wide flange T beams. In such a system the simple engineering beam theory assumption that sections that are plane before bending occurs remain plane thereafter is no longer justified. There is considerable distortion of the plate, as illustrated in

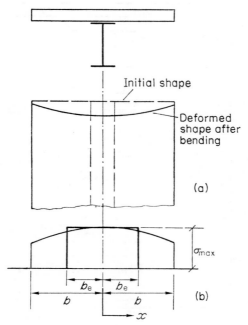

Figure 2.7 Shear lag in wide flange beam

Figure 2.7, the distortion being accompanied by a non-uniform longitudinal stress distribution across the section (Figure 2.7(b)). Theoretical solutions for stress distribution exist but they are complex and ill-adapted for design use as they are dependent not only on the relative dimensions and stiffnesses of the system but also on the nature of the applied loading. In addition the stress distribution to be found at any particular section will not generally be repeated at any other section. Because the plate distortion is a consequence of varying bending stress, and hence of shear stress, the phenomenon is known as *shear lag*.

The concept of effective width is generally adopted as a simplification of the shear lag problem. The effective width b_e is defined as 'that width of slab which acted on by the actual maximum stress would cause the same static effect as the variable stress which exists in fact'.

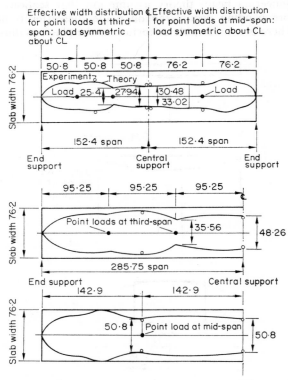

Figure 2.8 Effective width distributions in small scale beams with point loads at mid-span and third spans. (Dimensions in millimetres. Theoretical solutions ignoring slip)[5]

Referring to Figure 2.7(b)

$$2b \int_0^b \sigma_x \mathrm{d}x = 2b_e \sigma_{max}$$

where σ_{max} is the maximum value of stress in the slab and σ_x is the value of stress at a distance x from the origin.

Experimental work on the distribution of effective width has been carried out,[5] a typical plot being shown in Figure 2.8. When the results are compared with theoretical solutions it is generally found that the experimental values are greater. Taking into account that the theoretical solutions ignore inelastic behaviour and transverse cracking, and that in any case variations in effective width have a relatively small effect on the steel and concrete stresses, there is little to be gained by any complicated derivation of effective width.

The CEB Recommendations[2] include comprehensive rules and tables for determining effective widths taking into account the following factors;

(a) simply supported or continuous beams
(b) concentrated or distributed loading
(c) the ratio of flange thickness to beam depth
(d) the ratio of length of beam between points of zero moment to the width of the web and to the distance between webs

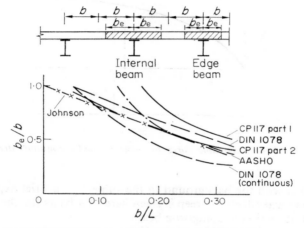

Figure 2.9 Comparison of effective width formulae.

For many purposes a simpler rule restricting the effective width to some proportion of the beam span (L) is all that is required; the CEB Recommendations suggest $2b_e = L/8$ for uniformly distributed loads and $2b_e = L/10$ for point loads, where L is the distance between points of zero bending moment and $2b_e$ is the *total* effective width of the flange.

CP 117 part 2[6] has the formula

$$b_e/b = 1/\left[1 + 12\,(2b/L)^2\right]^{\frac{1}{2}}$$

which is valid when b exceeds 0·05 of the span. Below this value $b_e = b$, above it b_e is reduced in accordance with the formula but not below 0·05 of the span.

Based on ultimate load conditions, Johnson[7] has proposed $b_e = (1-b/L)b$. Figure 2.9 gives a comparison of a number of ways of computing the effective width.

2.4 Creep

Theoretical and experimental studies of the nature of creep (plastic strain occurring under load in concrete) have been widely reported[8]. The reader is therefore referred to the many sources of information

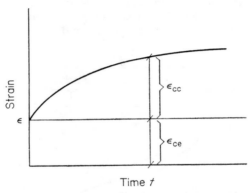

Figure 2.10 Strain–time curve for concrete under constant stress

which deal with the background to the subject — a brief explanation of the physical effects of creep is given here as a basis for the consideration of its action in composite beams.

When concrete sustains stress there is an instantaneous elastic strain, followed by a time-dependent increase in strain known as creep. Conversely if the strain in concrete does not alter with time the stress will reduce, a process known as *relaxation*. If for example, a concrete specimen carries a constant load applied at some time after the concrete has been made the general form of the strain–time curve will be as shown in Figure 2.10. The value of the elastic strain ε_{ce} is calculated in the conventional way using the *instantaneous modulus of elasticity* appropriate to the concrete at the age of loading. The value of the plastic, creep strain ε_{cc} increases with time; at any given time after loading the *apparent modulus of elasticity* will be given by the expression $E_t = \text{stress}/(\varepsilon_{ce} + \varepsilon_{cc})$.

The instantaneous modulus of elasticity is not a constant quantity (Figure 2.11); its value varies with the concrete strength and, because concrete strength increases with time, the instantaneous modulus does the same. However, it is accepted that little inaccuracy arises from neglect of the increase of the instantaneous modulus.

A *creep coefficient* ϕ is defined as the ratio of creep strain to elastic strain:

$$\phi = \varepsilon_{cc}/\varepsilon_{ce}$$

The value of the creep coefficient is a function of the time which has elapsed since the commencement of loading and we may write its value at time t as

$$\phi_t = k_c k_d k_b k_e k_t$$

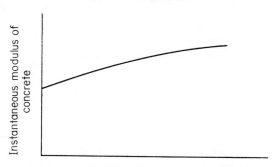

Figure 2.11 *Increase in instantaneous modulus of elasticity of concrete with age*

where the constants k are determined from the following considerations:

    k_c environmental
    k_d age of concrete at commencement of loading
    k_b the composition of concrete
    k_e the thickness of concrete
    k_t the elapsed time since commencement of loading

Curves giving values for the five constants k are plotted in Figure 2.12.[2]

Thus under constant stress σ the total strain in concrete at time t after loading is

$$(\sigma/E)(1+\phi_t)$$

However, when a variable stress acts on the concrete, calculation of the strain at any given time becomes much more complicated. Where stresses are well below the ultimate strength of the concrete it is possible to superpose the effects of each stress change, on the assumption that they do not interact. Given knowledge of the creep–time curve appropriate to the time at which each stress change occurs,

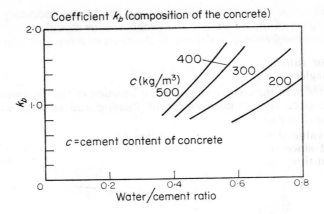

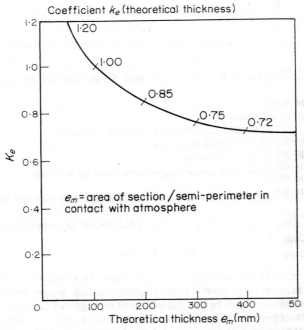

Figure 2.12 Coefficients for calculating creep coefficient ϕ_t (see reference 2).
area of semi perimeter in

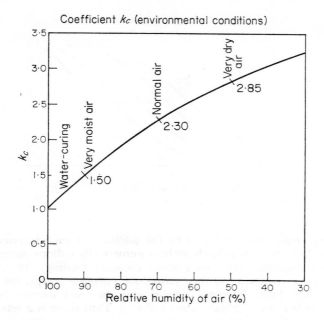

Coefficient k_c (environmental conditions)

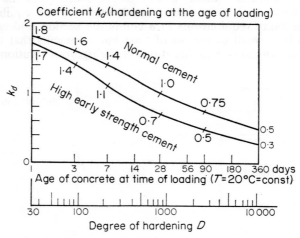

Coefficient k_d (hardening at the age of loading)

c = *cement content (kg) per cubic metre of concrete.* e_m = *theoretical thickness = contact with atmosphere*

Coefficient k_f (variation as a function of time)

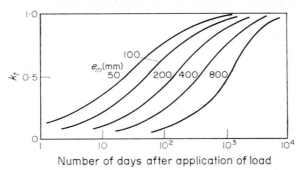

Number of days after application of load

Figure 2.12 continued

the total strain may be found by the addition of each increment. It will be clear that this is likely to be a numerically tedious process.

In a composite beam, even under constant loading, the concrete stress is not constant; as time passes the concrete creeps, the strain in the steel beam becomes greater and the steel stresses become larger while those in concrete are reduced. Thus there is a relaxation of stress in the concrete, the value of which is connected with the relative stiffnesses of beam and slab. The stress changes are shown in Figure 2.13.

The calculation of creep stresses in composite beams is theoretically very complicated. Sattler[9] gives rigorous solutions of the differential equations which arise from consideration of the equilibrium and compatibility requirements in a composite section undergoing creep strain in a small increment of time but comments that their use in practice is tedious. Simplifications of the rigorous solutions and other

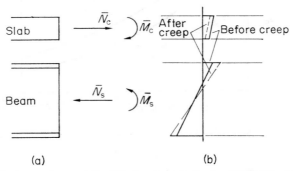

Figure 2.13 Stress changes caused by creep. (a) Section force changes caused by creep; (b) stresses before and after creep has occurred

methods are outlined in section 2.7 In many situations the stress changes caused by creep are small and can be ignored in practical design. Alternatively a simple way of allowing for them is to use an effective modulus of elasticity as described in section 2.7.1.

2.5 General Analysis of a composite section

A rigorous general analysis of the composite section demands a complete knowledge of the full stress–strain properties of the three components, steel, concrete and shear connection, over the time span being considered. Theoretical studies have been carried out on such a basis[3] but the complex numerical work involved makes the use of a computer essential, and for practical purposes some less cumbersome method is required. By making simplifying assumptions about the properties of the component materials the numerical work can be eased without significant loss of accuracy. It is instructive first to

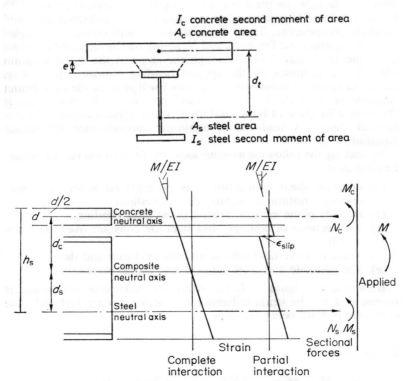

Figure 2.14 Strain diagrams for complete and partial interaction

treat the case of 'partial interaction' where there is relative movement between slab and beam caused by deformation of the shear connection. Although in practice this movement (slip) may have little significance, its inclusion in the analysis produces a measure of its importance.

2.5.1 PARTIAL INTERACTION

When account is taken of the deformation of the shear connection between beam and slab there will be a loss of bending stiffness in the composite beam which will become greater the more flexible the shear connection, reaching a limit when interaction ceases and beam and slab act independently. Theories of partial interaction have been developed and are based on greater or lesser simplifying assumptions. The one presented here is due to Newmark.[10]

It is convenient to consider the bending moment (and axial forces due, for example, to prestressing) acting on the composite beam to be the resultant of *sectional forces* acting on the individual steel and concrete components. These sectional forces will consist of couples M_s and M_c and axial forces N_s and N_c acting on the steel and concrete components (Figure 2.14) Although the slope of the strain diagram (M/EI) is determined by the applied bending moment, there is an infinity of values of the slip strain ε_{slip} which will produce the equilibrium requirement that the forces N should be equal. The slip strain is therefore a function of both axial force N and applied moment M and general treatment leads to a nonlinear second-order differential equation.[3]

By making the following assumptions, the form of the equation may be simplified:

(a) Discrete shear connection may be replaced by an equivalent uniform continuous linearly elastic medium
(b) Initially plane sections remain plane after bending
(c) The sections of concrete and steel are constant over the beam length
(d) There is no vertical separation between beam and slab
(e) The concrete and steel are isotropic elastic materials

Referring to Figure 2.14, the effect of incomplete interaction is represented by the strain difference ε_{slip} between beam and slab, the curvature in both being the same.

For equilibrium

$$M = M_c + M_s + N d_t \qquad (N = N_c = -N_s) \qquad (2.1)$$

For compatibility

$$M_c/E_cI_c = M_s/E_sI_s \quad \text{(equal curvatures)} \quad (2.2)$$

Consider a portion of composite beam dx long. If the slip is ds then slip strain is given by

$$\varepsilon_{\text{slip}} = ds/dx \quad (2.3)$$

Now the slip in the connecting medium is caused by the interface shear dN/dx.

Let the modulus of the medium (slip caused by unit shear) be μ then the slip in length dx

$$ds = \mu(dN/dx)$$

$$\varepsilon_{\text{slip}} = \mu(d^2N/dx^2) \quad (2.4)$$

The slip strain $\varepsilon_{\text{slip}}$ is caused by the difference between the strains in the slab and beam at the interface.

Strain in slab at interface $= -N/A_cE_c+(M_c/E_cI_c)(d_i/2+e) \quad (2.5)$

Strain in beam at interface $= N/E_sA_s-(M_s/E_sI_s)[d_t-(d/2)-e] \quad (2.6)$
(tensile strain positive).

Substituting for M_c in (2.5) from (2.2) and subtracting (2.6) from (2.5),

$$N(1/A_sE_s+1/A_cE_c)-M_sd_t/E_sI_s = \varepsilon_{\text{slip}} = \mu \, d^2N/dx^2 \quad (2.7)$$

From (2.1) and (2.2)
$$M_s = (M-Nd_t)\{(E_sI_s/(E_cI_c+E_sI_s)\} \quad (2.8)$$

For simplicity the various sums and products of the section constants may be written:

$$E_cA_cE_sA_s/(E_cA_c+E_sA_s) = \overline{EA} \quad (2.9)$$

$$E_cI_c+E_sI_s = \Sigma EI \quad (2.10)$$

$$\Sigma EI+\overline{EA} \, d_t^2 = \overline{EI} \quad (2.11)$$

The final form of the equation, after the elimination of M_s, becomes

$$\mu(d^2N/dx^2)-N \, \overline{EI}/\overline{EA} \, \Sigma EI+Md_t/\Sigma EI = 0 \quad (2.12)$$

or

$$d^2N/dx^2-B^2N+C = 0 \quad (2.13)$$

where

$$B^2 = \overline{EI}/\mu\Sigma EI \, \overline{EA} \quad (2.14)$$

$$C = Md_t/\mu\Sigma EI \quad (2.15)$$

3*

The solution of equation (2.13) will give the distribution of the axial force N and the interface shear dN/dx for various boundary conditions and loading cases for given values of shear connection constant μ.

Figure 2.15 contains solutions for the axial force N and interface shear produced by uniform load, point load and temperature difference in terms of N_o and v_o which result from rigid shear connection.[11] The solution for temperature difference (which is also the solution for shrinkage) is of particular interest as it discloses the existence of high interface shear at the beam ends with inflexible connection. Putting $x = 0$,

$$v = N_o[(B \sinh B L/2)/(\cosh BL/2)] = N_o B \tanh BL/2$$

For rigid shear connection (μ very small) B is very large and

$$\tanh BL/2 \to 1{\cdot}0$$

v then becomes very large and for absolutely rigid shear connection the interface shear at the ends of a simply supported beam due to differential shrinkage or temperature strain is theoretically infinite.

2.6 Elastic analysis

Elastic design methods require that stresses in the beam must nowhere exceed specified working stress values, these values being based on steel yield or concrete crushing stresses reduced by a suitable factor of safety.

2.6.1 TRANSFORMED AREA METHOD

For an elastic analysis the following assumptions are generally made:

(a) The shear connection between beam and slab is sufficient to ensure that slip does not significantly affect the assumption of full interaction
(b) Both steel and concrete are linearly elastic materials
(c) Concrete undergoing tensile strain is ineffective in resisting load

The *transformed area* method can usefully be employed so that the composite beam can be treated as if composed of a single material. This effect is achieved by dividing the actual concrete area (the product of effective width and depth in *compression*) by the relevant modular ratio $m = E_s/E_c$. In cases where the steel beam is relatively small in depth in relation to the slab, the composite neutral axis may

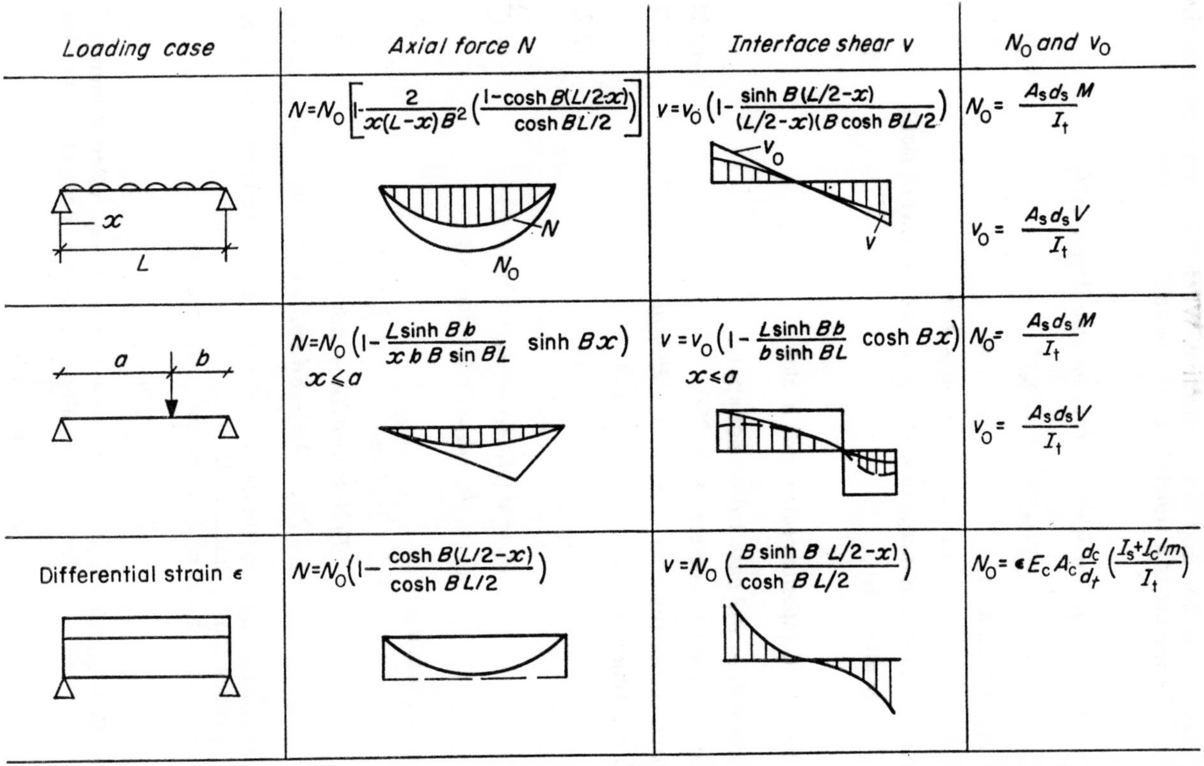

Loading case	Axial force N	Interface shear v	N_0 and v_0
	$N = N_0\left[1 - \dfrac{2}{x(L-x)B^2}\left(\dfrac{1-\cosh B(L/2-x)}{\cosh BL/2}\right)\right]$	$v = v_0\left(1 - \dfrac{\sinh B(L/2-x)}{(L/2-x)(B\cosh BL/2)}\right)$	$N_0 = \dfrac{A_s d_s M}{I_t}$ $v_0 = \dfrac{A_s d_s V}{I_t}$
	$N = N_0\left(1 - \dfrac{L\sinh Bb}{x\,b\,B\sin BL}\sinh Bx\right)$ $x \le a$	$v = v_0\left(1 - \dfrac{L\sinh Bb}{b\sinh BL}\cosh Bx\right)$ $x \le a$	$N_0 = \dfrac{A_s d_s M}{I_t}$ $v_0 = \dfrac{A_s d_s V}{I_t}$
Differential strain ϵ	$N = N_0\left(1 - \dfrac{\cosh B(L/2-x)}{\cosh BL/2}\right)$	$v = N_0\left(\dfrac{B\sinh B(L/2-x)}{\cosh BL/2}\right)$	$N_0 = \epsilon E_c A_c \dfrac{d_c}{d_f}\left(\dfrac{I_s + I_c/m}{I_t}\right)$

Figure 2.15

fall within the slab. The concrete below the neutral axis is then cracked in tension and thus ineffective. When this occurs the effective area of concrete is a function of the neutral axis position.

Case 1 — Composite neutral axis below slab (Figure 2.14)

Actual area of concrete	$A_c = bd$	(2.16)
Transformed area	$= A_c/m$	(2.17)
Total composite area	$A_t = A_s + A_c/m$	(2.18)
$A_s d_t = A_t d_c$ (moments of area about neutral axis of slab)		(2.19)
$d_c = d_t(A_s/A_t)$		(2.20)

Case 2 — Composite neutral axis in slab

Where $d_c < d/2$ case 1 does not apply.

Actual area of concrete $A_c = 2bd_c$ (2.21)

Total composite area $A_t = A_s + 2bd_c/m$ (2.22)

d_t is a function of neutral axis depth.

$$d_t = h_s - d_c \tag{2.23}$$

$$A_s d_t = A_t d_c$$

$$d_c = d_t(A_s/A_t) = (h_s - d_c)(A_s/A_t) \tag{2.24}$$

$$= (h_s - d_c)A_s/(A_s + 2bd_c/m)$$

or, rearranging,

$$A_s d_c + 2bd_c^2/m - A_s h + A_s d_c = 0$$

$$d_c^2(2b/m) + d_c 2A_s - A_s h = 0 \tag{2.25}$$

$$d_c = -2A_s \pm (4A_s^2 - 8A_s bh_s/m)^{1/2}/(4b/m) \tag{2.26}$$

Having determined the depth of the neutral axis the remaining section properties can be calculated.

Second moment of area of composite section

$$I_t = I_s + I_c/m + A_c d_c^2/m + A_s d_s^2$$

Put $S_t = A_c d_c/m = A_s d_s$ (moment of steel or equivalent concrete area about composite neutral axis)

$$I_t = I_s + I_c/m + S_t d_c + S_t d_s$$

$$= I_s + I_c/m + S_t d_t \tag{2.27}$$

The stress at any level can then be computed from $\sigma = My/I_t$ bearing in mind that concrete stresses are obtained from $\sigma = My/mI_t$.

Axial forces may act on a composite section from, for example, prestressing cables in the slab. The stresses caused by an axial force $\bar{N}$ acting at an arbitrary eccentricity e from the composite neutral axis are found by considering the statically equivalent system:

An axial force $\bar{N}$ at the composite centroid plus a bending moment $\bar{N}e$.

Stresses due to the moment $\bar{N}e$ are calculated as for any bending moment on the section

Stresses due to $\bar{N}$:

$$\sigma_c = \bar{N}/mA_t \qquad \sigma_s = \bar{N}/A_t$$

2.6.2 DISTRIBUTION OF BENDING MOMENT AND AXIAL FORCE

It is useful, in analysing the influence of time-dependent effects, to distribute applied moments and axial forces between steel and concrete in the manner already demonstrated for partial interaction. The expressions below are obtained from those for partial interaction (see section 2.5.1) by putting the slip strain ε_{slip} equal to zero. (The distributed bending moments and axial forces are called 'sectional forces' in this book.)

$$M_c = M(I_c/mI_t) \tag{2.28}$$

$$M_s = M(I_s/I_t) \tag{2.29}$$

$$N_c = -N_s = M(d_cA_c/mI_t) = M(S_t/I_t) \tag{2.30}$$

It should be noted that because the concrete slab is often relatively small ($I_c/m \ll I_s$), M_c may, in such cases, be neglected.

Horizontal shear forces v at the beam–slab interface, due to a vertical shear force V acting at any section, may be obtained from

$$V = dM/dx \tag{2.31}$$

$$v = dN/dx \tag{2.32}$$

From 2.30

$$dN/dx = (S_t/I_t)dM/dx = (S_t/I_t)V$$

$$v = (S_t/I_t)V \tag{2.33}$$

2.7 Calculation of stress changes due to creep

It has been pointed out in section 2.4 that rigorous solutions of the creep problem are difficult to use in practice and so approximate methods have been developed in order to simplify calculation of creep effects. Very often these effects are small enough to be neglected; in fact some codes of practice specifically do not require them to be evaluated for simply supported beams. Indeterminate structures, on the other hand, may develop significant creep stresses.

In the sections which follow, some methods of calculation are examined. The notation adopted is as follows.

Initial section forces at instant of loading

$$M_c, \qquad M_s, \qquad N = N_c = -N_s$$

Sectional forces at time t after loading

$$M_{ct}, \qquad M_{st}, \qquad N_t = N_{ct} = -N_{st}$$

The *change* in sectional force in a time interval t is shown by a bar thus

$$\bar{N}_s = N_{st} - N_s = \bar{N}_t \qquad N_c = N_{ct} - N_c = -\bar{N}_t$$
$$\bar{M}_c = M_{ct} - M_c$$
$$\bar{M}_s = M_{st} - M_s$$

Indeterminate structures are treated separately in section 2.9.

2.7.1 EFFECTIVE MODULUS METHOD

A simple way of evaluating the effect of long term loading is to adopt a modulus of elasticity appropriate to the duration of load. Whereas, for very short loading times, the initial tangent modulus E_c is adopted, for longer times E_c is modified to take account of creep, i.e. the effective (apparent) modulus is used. Concrete structures are commonly designed in this way; in fact a single modulus is often used which is approximately the mean of the instantaneous and apparent modulus at infinite time. The value of modular ratio $m = 15$ has a long, respected history in reinforced concrete theory; it corresponds to a single effective concrete modulus of $E_c = 1.4 \times 10^4$ N/mm². This treatment of concrete as a linearly elastic medium, although it does not accord with the physical nature of the material, does lead to ease of design.

In practice, loads are considered to be either of very short or infinite duration. The two moduli, E_c (short term) and E'_c (infinite), are related by $E'_c = E_c/k$ where the factor k lies generally between 2 and 4 and may have a fixed value or be related to the final creep coefficient. Some values of k are given in *Table 2.3*.

Table 2.3 VALUES OF EFFECTIVE
MODULUS FACTOR

Source	k
CP 117 part 2	2
AASHO	3
Dischinger[9]	$(1+\phi_N)$
Fritz[9]	$(1\cdot1+\phi_N)$

Note: ϕ_N = final creep coefficient. CP 117: Code of Practice
117 (BSI)

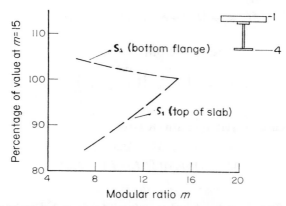

Figure 2.16 Change in section modulus S for changes in modular ratio m

The effect of variations in the value of the modular ratio on the
section properties of a composite beam are shown graphically in
Figure 2.16 from which it will be seen that there is little change in
the bottom flange section modulus over a range of *m* from 7 to 15.
As the bottom flange stresses are generally critical in design, the
neglect of creep stresses in statically determinate beams appears to
be justified.

2.7.2 APPROXIMATE METHOD DUE TO SATTLER[9]

Rigorous solutions, due to Sattler, are based on Dischinger's equation
for the strain in concrete that undergoes a stress change during a small
time increment:

$$d\varepsilon_c/dt = (1/E_c)\,(d\sigma_c/dt)+(\sigma_c/E_c)d\phi/dt \qquad (2.34)$$

The differential equations resulting from the adoption of equation (2.34) lead to complicated solutions, ill adapted for practical use. Sattler therefore proposes approximate relationships.[12] There are two cases to be considered, distinguished by the value of the parameter

$$\lambda = A_c I_c / m^2 A_s I_s$$

which is a measure of the relative size of the slab to the beam.

(a) $\lambda \leqslant 0\cdot2$

In this case the slab is relatively small and both M_c and $\overline{M}_c$ are small in relation to M_s.

$$\overline{N} = N(1 - e^{-\alpha_s\phi}) \qquad (2.35)$$

$$\overline{M}_s = \overline{N}_s d_t \qquad (2.36)$$

$$\overline{M}_c = -M_c(1 - e^{-\phi}) + (I_c/mI_s)\psi\,\overline{M}_s \qquad (2.37)$$

where

$$\alpha_s = A_s I_s / A_t I_t \qquad (2.38)$$

$$\psi = \left(\frac{\alpha_s}{1-\alpha_s}\right)\left(\frac{e^{\alpha_s\phi} - e^{-\phi}}{1 - e^{-\alpha_s\phi}}\right) \qquad (2.39)$$

(b) $\lambda > 0\cdot2$

In this case the slab is relatively large.

$$\overline{N}_c = (1/d_t)\,(\overline{M}_s + \overline{M}_c) \qquad (2.40)$$

$$\overline{M}_c = (1/\Phi_s)\,[(\overline{M}_s I_c/mI_s) - M_c\phi] \qquad (2.41)$$

$$\overline{M}_s = \frac{-d_t\phi N + M_c\phi/\Phi_s(\Phi_s + A_c/mA_s)}{\Phi_s + A_c/mA_s + I_c/mI_s + A_c I_s/m^2 A_s I_s \Phi_s + A_c d_t^2/mI_s} \qquad (2.42)$$

where $\Phi_s = (1 + 0\cdot65\phi)$. Tabular values of $(1 - e^{-\phi})$, $(1 - e^{-\alpha_s\phi})$ and ψ are available.

2.7.3 RELAXATION METHOD

A method which is suitable for hand calculation has been proposed by Trost,[13] the basis of which is an equation for the strain in concrete at time t after loading:

$$\varepsilon_{ct} = (\sigma_{co}/E_c)\,(1 + \phi_t) + [(\sigma_{ct} - \sigma_{co})/E_c]\,(1 + \varrho\phi_t) \qquad (2.43)$$

where σ_{co} is the initial stress at time t_o, σ_{ct} is the stress at time t and ϱ is a relaxation coefficient which takes account of the reduced creep tendency of stress changes occurring at times after t_o.

The adoption of equation (2.43) enables the values of sectional forces M_{st}, M_{ct} and N_t to be calculated from considerations of equilibrium and compatibility in the composite beam. Unlike the solutions using Dischinger's expression for the strain in a small element of time, which lead to complicated differential equations, the relaxation solution leads directly to simultaneous equations.

Values of ϱ, the relaxation coefficient, are given in Figure 2.17, as a function of the age of concrete at first loading, the final creep coeffi-

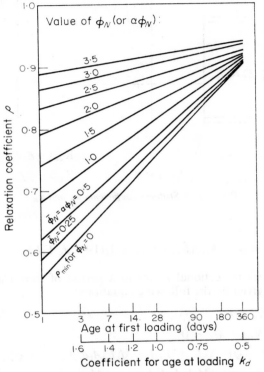

Figure 2.17 Values of relaxation coefficient ϱ as a function of age at first loading[8]

cient ϕ_N and two stiffness coefficients α_N and α_M. These last two coefficients reflect the fact that creep is affected by the restraining action of the steel beam.

The axial force stiffness coefficient α_N is defined by the action of unit axial force acting at the concrete centroid (Figure 2.18).

$$\alpha_N = \varepsilon_{co}(\varepsilon_{co} + \varepsilon_{so}) \tag{2.44}$$

from which

$$\alpha_N = 1/[1 + (E_c A_c / E_s A_s)(1 + A_s d_t^2 / I_s)] \tag{2.45}$$

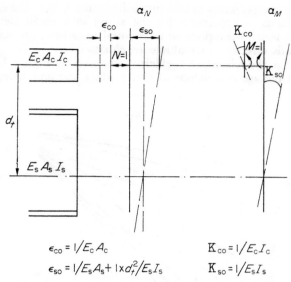

$$\epsilon_{co} = 1/E_c A_c \qquad\qquad K_{co} = 1/E_c I_c$$
$$\epsilon_{so} = 1/E_s A_s + 1 \times d_t^2/E_s I_s \qquad K_{so} = 1/E_s I_s$$

Figure 2.18 Stiffness coefficients α_N and α_M

Similarly

$$\alpha_M = K_{co}/(K_{co}+K_{so}) = 1/(1+E_c I_c/E_s I_s) \tag{2.46}$$

The changes in sectional forces in a period of time t after initial loading are given by the following equations:

$$\bar{N}_c + \bar{N}_s = 0 \tag{2.47}$$

$$\bar{M}_s + \bar{N}_s d_t + \bar{M}_c = 0 \tag{2.48}$$

$$\bar{N}_c(1+\varrho_N\phi) - \bar{N}_s E_c A_c/E_s A_s + \bar{M}_c E_c A_c d_t/E_s I_s = -\phi N_c \tag{2.49}$$

$$\bar{M}_c(1+\varrho_M\phi) - M_s E_c I_c/E_s I_s = -\phi M_c \tag{2.50}$$

The four unknown sectional forces $\bar{N}_c$, $\bar{N}_s$, $\bar{M}_c$, $\bar{M}_s$ may be found from the solution of these four equations.

Substituting from (2.47) and (2.48) into (2.49) and (2.50) and using the stiffness ratios gives the following equations in terms of concrete section forces:

$$\bar{N}_c(1+\varrho_N\alpha_N\phi) - \frac{\bar{M}_c\alpha_N E_c A_c A_s d_t^2}{d_t E_s A_s I_s} = -\alpha_N\phi N_c \tag{2.51}$$

$$\bar{N}_c(1-\alpha_M) - \frac{\bar{M}_c}{d_t}(1+\varrho_M\alpha_M\phi) = \alpha_M\phi\frac{M_c}{d_t} \tag{2.52}$$

or in terms of steel sectional forces (without stiffness ratios):

$$\bar{N}_s(1 + E_cA_c/E_sA_s + \varrho_N\phi) - \frac{\bar{M}_s}{d_t}\frac{E_cA_c}{E_sA_s}\frac{A_sd_t^2}{I_s} = \phi N_c \qquad (2.53)$$

$$\bar{N}_s(1 + \varrho_M\phi) + \frac{M_s}{d_t}(1 + E_cI_c/E_sI_s + \varrho_M\phi) = \phi M_c/d_t \qquad (2.54)$$

The calculation of the unknown sectional forces is therefore reduced to the solution of two equations in two unknowns.

Commonly the value of $\bar{M}_c$ is small in relation to $\bar{M}_s$ and so may be neglected (the criterion $I_c/mI_s < 0.05$) reducing equation (2.48) to $\bar{M}_s + \bar{N}_s d_t = 0$. Equation (2.51) then becomes

$$\bar{N}_c(1 + \varrho_N\alpha_N\phi) = -\alpha_N\phi N_c \qquad (2.55)$$

and $\bar{M}_s$ may be calculated from

$$\bar{M}_s = -\bar{N}_s d_t = \bar{N}_c d_t \qquad (2.56)$$

Rearranging equation (2.52)

$$\bar{M}_c = \frac{\bar{N}_c(1 - \alpha_M)d_t}{(1 + \varrho_M\alpha_M\phi)} - \frac{M_c\alpha_M\phi}{(1 + \varrho_M\alpha_M\phi)} = \frac{d_t - \bar{N}_c(1 - \alpha_M) - M_c\alpha_M\phi}{(1 + \varrho_M\alpha_M\phi)} \qquad (2.57)$$

2.7.4 CREEP FIBRE METHOD

In a composite beam there are pairs of conjugate points which have the property that an axial force applied at one point of any pair causes zero stress at the other.

In general, referring to Figure 2.19, for an axial compressive force N applied at a distance e below the composite neutral axis, there will be a point y_o above the composite neutral axis at which the resultant of axial and bending stress is zero.

$$N/A_t = (Ne/I_t)\,y_0$$

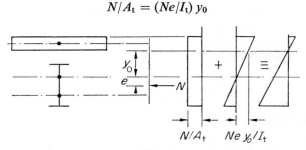

Figure 2.19 Location of conjugate zero stress points

or

$$y_o = (I_t/A_t)(1/e) \tag{2.58}$$

The action of creep or shrinkage may be represented by a resultant force acting at some level in the concrete slab. By using this level as one point of a pair of conjugate 'zero stress' points the position of the other may be found; these two particular points, known as *creep fibres*, may then be used to determine creep stresses as if they represented the centroids of two axially loaded columns. Because of the independence of the creep fibres, loads applied to one of them do not affect the other, and so each point may be considered separately and the effects superimposed.[14]

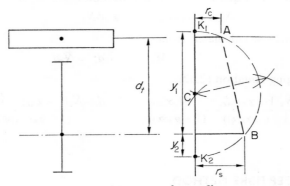

Figure 2.20 Location of creep fibres

A geometrical construction may be used to locate the creep fibres (Figure 2.20). At the concrete and steel centroids, lengths r_c and r_s are drawn where

$$r_c = (I_c/A_c)^{1/2}$$
$$r_s = (I_s/A_s)^{1/2}$$

The line AB is bisected to locate point C from which a circle passing through A and B is drawn. The points K_1 and K_2 are the two creep fibres.

From the geometry of the diagram

$$y_1 y_2 = r_s^2 \tag{2.59}$$
$$(d_t + y_2)(y_1 - d_t) = r_c^2 \tag{2.60}$$

From which

$$y_2 = \frac{r_s^2 - d_t^2 - r_c^2}{2d_t} \pm \left(r_s^2 + \frac{(r_s^2 - d_t^2 - r_c^2)^2}{2d_t} \right)^{1/2} \tag{2.61}$$

In practice it will be found that K_1 is almost coincident with the centroid of the concrete because r_c is small when compared with r_s. From equation (2.61), putting $r_c = 0$ and taking the positive root,

$$y_2 = 2r_s^2/2d_t = I_s/A_s d_t \tag{2.62}$$

The same result may be obtained from equation (2.58) by writing $y_o = d_c$ and $e = y_2 + d_s$:

$$y_2 = (I_t/A_t d_c) - d_s \tag{2.63}$$

$$I_t = I_s + A_s d_s d_t \quad \text{(neglecting } I_c/m)$$

$$A_t d_c = A_s d_t$$

$$y_2 = (I_s + A_s d_s d_t - A_s d_s d_t)/A_s d_t = I_s/A_s d_t \tag{2.64}$$

Thus, unless the concrete area is large in relation to the steel beam area, the creep fibres may be assumed to be located at the centroid of the slab and at a point y_2 below the centroid of the steel beam defined by equation (2.63)

Having found the creep fibres, the areas of the substitute columns with centroids K_1 and K_2 are

$$A_{s1} = A_s[y_2/(y_1 + y_2)] \qquad A_{c1} = A_c[(y_2 + d_t/(y_1 + y_2)] \tag{2.64a}$$

$$A_{s2} = A_s[y_1/(y_1 + y_2)] \qquad A_{c2} = A_c[(y_1 - d_t)/(y_1 + y_2)] \tag{2.65}$$

or once again if K_1 is coincident with the slab centroid, putting $y_1 = d_t$ the concrete areas become

$$A_{c1} = A_c$$

$$A_{c2} = A_c(d_t - d_t)/(d_t + y_2) = 0$$

In its simplified form, then, the problem resolves itself into the evaluation of the stress change in the creep fibre K_1 caused by an axial force acting on a reinforced concrete column of concrete area A_{c1} and steel area A_{s1}, the axial force being found from statical consideration of the internal and external forces and moments acting on the section. This may be done by using a relaxation technique or by adopting Dischinger's differential equation (see Example 2.4 p. 42)

2.7.5 EXAMPLES OF THE CALCULATION OF CREEP STRESS

The examples which which follow illustrate the calculation, by various methods, of the stresses due to creep in the composite beam of Figure 2.21.

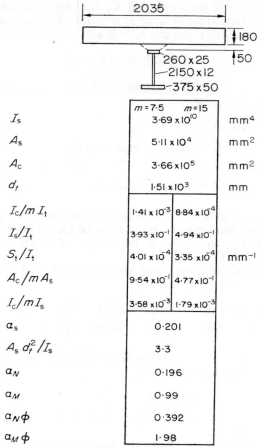

Figure 2.21 *Properties of composite beam. (Dimensions in mm)*

Final creep coefficient $\phi_N = 2{\cdot}0$.   Tension — positive
Initial bending moment 4035 kN m. Sagging bending — positive

Sectional forces and moments at time of loading (t = 0, m = 7.5)

$$N = (S_t/I_t)M \qquad (4035 \times 4{\cdot}01 \times 10^{-1}) = +1620 \text{ kN}$$
$$N_c = -N \qquad\qquad\qquad\qquad\qquad = -1620 \text{ kN}$$
$$N_s = +N \qquad\qquad\qquad\qquad\qquad = +1620 \text{ kN}$$
$$M_c = (I_c/mI_t)M \qquad (4035 \times 1{\cdot}41 \times 10^{-3}) = +5{\cdot}68 \text{ kN m}$$
$$M_s = (I_s/I_t)M \qquad (4035 \times 3{\cdot}93 \times 10^{-1}) = +1590 \text{ kNm}$$

The smallness of M_c in relation to M_s should be noted. The change in the sectional forces due to creep will now be calculated by the effective modulus method, Sattler's approximate method and the relaxation method.

EXAMPLE 2.1 EFFECTIVE MODULUS METHOD

With $m = 15$ the sectional forces and moments are

$$N_t = 1.35 \times 10^3 \text{ kN}$$
$$N_{ct} = -1.35 \times 10^3 \text{ kN}$$
$$N_{st} = +1.35 \times 10^3 \text{ kN}$$
$$M_{ct} = 3.57 \text{ kN m}$$
$$M_{st} = 2.01 \times 10^3 \text{ kN m}$$

and the changes are

$$\bar{N}_t = (1.62 - 1.35) \times 10^3 = +270 \text{ kN}$$
$$\bar{N}_c = +270 \text{ kN}$$
$$\bar{N}_s = -270 \text{ kN}$$
$$\bar{M}_c = -(5.68 - 3.57) = -2.11 \text{ kN m}$$
$$\bar{M}_s = -(-1.59 - 2.01) \times 10^3 = +420 \text{ kN m}$$

EXAMPLE 2.2 SATTLER'S APPROXIMATE METHOD

The required parameters are

$$(1 - e^{-\alpha_s \phi}) = (1 - 0.67) = 0.33$$
$$(1 - e^{-\phi}) = (1 - 0.136) = 0.864$$
$$\psi = [\alpha_s/(1 - \alpha_s)] [(e^{-\alpha_s \phi} - e^{-\phi})/(1 - e^{-\alpha_s \phi})] = 0.405$$

whence $\bar{N}_t \ (= 1.62 \times 0.33 \times 10^3) = +540 \text{ kN}$

$$\bar{N}_c = +540 \text{ kN}$$
$$\bar{N}_s = -540 \text{ kN}$$
$$\bar{M}_c \ (= -5.68 \times 0.864 + 3.58 \times 0.405 \times 1.59) = -2.61 \text{ kN m}$$
$$\bar{M}_s \ (= 540 \times 1.51 \times 10^3) = 815 \text{ kN m}$$

EXAMPLE 2.3   RELAXATION METHOD

The dimensional parameters given in Figure 2.21 permit the relaxation characteristics to be obtained from Figure 2.17.

$$\varrho_N = 0.77 \qquad \varrho_M = 0.86$$

The numerical factors in equations (2.51) and (2.52) are

$$(1 + \varrho_N \alpha_N \phi)\, (= 1 + 0.77 \times 0.392) = 1.302$$

$$(1 + \varrho_M \alpha_M \phi)\, (= 1 + 0.86 \times 1.98) = 2.7$$

$$\alpha_N E_c A_c A_s d_t^2 / E_s A_s I_s d_t \,(= 0.196 \times 0.954 \times 3.3/1.51) = 0.408 \text{ m}^{-1}$$

$$(1 - \alpha_M)\, (= 1 - 0.99) = 0.01$$

$$(1 + \varrho_M \alpha_M \phi)/d_t \,(= 2.7/1.51) = 1.79 \text{ m}^{-1}$$

The approximate equation (2.55) gives $\bar{N}_c$ directly as

$$\bar{N}_c\, (= -(0.392/1.302)\,(-1620))\qquad = +487 \text{ kN}$$

$$\bar{N}_s \qquad\qquad\qquad\qquad\qquad\quad = -487 \text{ kN}$$

$$\bar{M}_s\, (= -(-487 \times 1.51))\; = +738 \text{ kN m}$$

$$\bar{M}_c \left(= \frac{487 \times 0.01}{1.51 \times 2.7} - \frac{5.68 \times 1.98}{2.7}\right) = -2.98 \text{ kN m}$$

It will be found that these values are almost identical with those obtained from solution of the exact equations.

If it is not desired to calculate the stiffness coefficients, a sufficiently accurate answer may be obtained by assuming $\alpha_N = 0$ and $\alpha_M = 1$, leading to relaxation characteristics $\varrho_N = 0.75$, $\varrho_M = 0.83$.

EXAMPLE 2.4   CREEP FIBRE METHOD

On the assumption that the upper creep fibre is located at the concrete centroid (inspection of the relative values of I_s and I_c will show this to be justified):

$$y_2 = I_s/A_s d_t \,(= 3.69 \times 10^{10}/5.11 \times 1.51 \times 10^7) = 4.78 \times 10^2 \text{ mm}$$

$$y_1 = d_t \qquad\qquad\qquad\qquad\qquad = 1.51 \times 10^3 \text{ mm}$$

$$y_1 + y_2 \qquad\qquad\qquad\qquad\qquad\;\; = 1.99 \times 10^3 \text{ mm}$$

$$A_{s1}\, (= 5.11 \times 10^4 \times 0.478/1.99) \quad = 1.23 \times 10^4 \text{ mm}^2$$

$$A_{c1} \qquad\qquad\qquad\qquad\qquad\qquad = 3.66 \times 10^5 \text{ mm}^2$$

$$A_{s2}\, (= 5.11 \times 10^4 \times 1.51/1.99) \quad = 3.88 \times 10^4 \text{ mm}^2$$

$$A_{c2} \qquad\qquad\qquad\qquad\qquad\qquad = 0$$

The applied bending moment of 4035 kN m may be considered as two equal and opposite forces of magnitude

$$(4035 \times 10^3/1 \cdot 99 \times 10^3) = 2030 \text{ kN}$$

acting at K_1 (compressive) and K_2 (tensile).

The calculation thus resolves itself into determining the creep stress on an axially loaded column of area $(A_{c1} + A_{s1})$ with centroid at K_1. The column centred on K_2 will not experience any creep because $A_{c2} = 0$.

Initial stresses

With $m = 7 \cdot 5$,

$A_1 (= 3 \cdot 66 \times 10^5 + 7 \cdot 5 \times 1 \cdot 23 \times 10^4) \quad = 4 \cdot 58 \times 10^5 \text{ mm}^2 \text{ (concrete)}$
$A_2 (= 3 \cdot 88 \times 7 \cdot 5 \times 10^4) \qquad\qquad = 2 \cdot 91 \times 10^5 \text{ mm}^2 \text{ (concrete)}$

giving initial stresses

at $K_1 (= 2 \cdot 03 \times 10^6/4 \cdot 58 \times 10^5) \quad = -4 \cdot 44 \text{ N/mm}^2 \text{ (concrete)}$
$\qquad\qquad\qquad\qquad\qquad\qquad\quad = -33 \cdot 1 \text{ N/mm}^2 \text{ (steel)}$
at $K_2 (= 2 \cdot 03 \times 10^6/2 \cdot 91 \times 10^5) \quad = +6 \cdot 84 \text{ N/mm}^2 \text{ (concrete)}$
$\qquad\qquad\qquad\qquad\qquad\qquad\quad = +51 \cdot 2 \text{ N/mm}^2 \text{ (steel)}$

Change in stresses due to creep

The change in force in the concrete of an axially loaded reinforced concrete column may be calculated using the relationship

$$\bar{N} = N(1 - e^{-\alpha\phi})/(1 + mp)$$

where
the proportion of reinforcement $p = 0 \cdot 123/3 \cdot 66 = 0 \cdot 0336$

$\alpha = mp(1 + mp) = 0 \cdot 0336 \times 7 \cdot 5(1 + 0 \cdot 0336 \times 7 \cdot 5) = 0 \cdot 20$
$\bar{N} = 2030 (1 - e^{-0 \cdot 2 \times 2})/(1 + 0 \cdot 0336 \times 7 \cdot 5)$
$\quad = +534 \text{ kN (tensile)}$

Thus the stress changes are

$(534 \times 10^3/3 \cdot 66 \times 10^5) = +1 \cdot 46 \text{ N/mm}^2 \text{ (concrete − tensile)}$
$(534 \times 10^3/1 \cdot 23 \times 10^4) = -43 \cdot 4 \text{ N/mm}^2 \text{ (steel — compressive)}$

4*

Table 2.4 COMPARISON OF METHODS

Method	Initial stresses				Change due to creep			
	f_1	f_2	f_3	f_4	$\bar{f}_1$	$\bar{f}_2$	$\bar{f}_3$	$\bar{f}_4$
Effective modulus	-4.95	-3.91	-27.3	$+68.5$	$+0.90$	$+0.58$	-21.1	$+4.6$
Sattler					$+1.72$	$+1.24$	-40.8	$+8.4$
Relaxation					$+1.60$	$+1.06$	-36.9	$+7.56$
Creep fibre					$+1.58$	$+1.42$	-40.2	$+8.15$

2.7.6 COMMENT ON CREEP CALCULATIONS

A comparison of the stress changes and final stress values for each of the four methods is given in *Table* 2.4. It will be observed that the methods produce comparable results, although the effective modulus method shows the smallest percentage change in stresses. At the critical point, the bottom flange, the increase in stress is 12 %; codes of practice commonly permit increases in basic allowable stress of more than 12 % for creep stresses. It does appear, then, that the effective modulus method is a simple and adequate method of calculating creep stresses.

2.8 Stresses caused by differential strain

Concrete shrinkage and temperature difference between beam and slab both lead to a restrained differential contraction or expansion between them. Referring to Figure 2.22 we may initially imagine that the slab is disconnected from the beam, thus being able to expand or contract without restraint. As a result of shrinkage or temperature difference the strain is ε.

To restore the slab to its initial length a force P will be imposed at the slab centroid

$$P = \varepsilon E_c A_c$$

At this stage the slab is reconnected to the beam.

The force P is now cancelled by an equal and opposite force P acting at the slab centroid (or by its static equivalent of a force P at the composite centroid and a couple $P\,d_c$).

Stresses in concrete and steel can now be computed (on the assumption that the slab initially shortens) — see bottom of p. 45.

OF CREEP STRESS CALCULATION

	Final stresses				*Ratio final/initial* (%)			
	f_1'	f_2'	f_3'	f_4'	f_1'/f_1	f_2'/f_2	f_3'/f_3	f_4'/f_4
Effective modulus	+4·05	−3·33	−48·4	+73·1	81·6	85·0	177	107
Sattler	−3·23	−2·67	−68·1	+76·9	65·3	68·1	250	112
Relaxation	−3·35	−2·85	−64·2	+76·1	67·5	72·9	235	111
Creep fibre	−3·37	−2·49	−67·5	+76·7	68·0	63·5	247	112

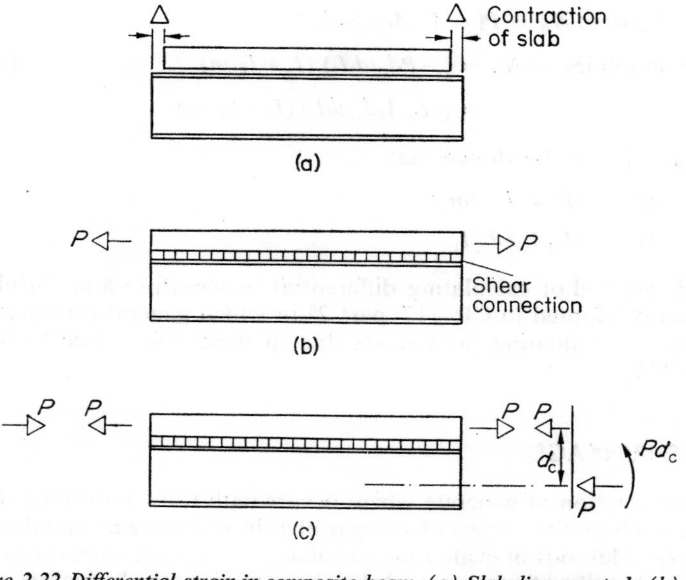

Figure 2.22 Differential strain in composite beam. (a) Slab disconnected; (b) Compatibility restored; (c) Equilibrium restored

Slab

$$\left.\begin{array}{ll} P/A_c & \text{tensile} \\ P/mA_t & \text{compressive} \end{array}\right\} + Pd_c y/mI_t$$

$$= P\left(1/A_c + 1/mA_t + d_c y/mI_t\right) \text{ compressive} \qquad (2.66)$$

Beam

$$(P/A_t) \text{ compressive} + Pd_c y/I_t = P(1/A_t + d_c y/I_t) \text{ compressive} \quad (2.67)$$

where the sign of y determines whether the bending stress is tensile or compressive. The magnitude of the stresses is calculated on the assumption of full interaction (zero slip).

The axial force N_o caused by differential strain may be found by applying the force P at the centroid of the composite section together with a bending moment $P\,d_c$ and summing the axial forces produced in the steel beam by these two effects:

Due to P:　$-P\,A_s/A_t$ compression

Due to $P\,d_c$:　$+Pd_c S_t/I_t$ tension

Total force:　$N_o = P(-A_s/A_t + S_t d_c/I_t)$

which simplifies to $N_o = (-Pd_c/d_t I_t)\,(I_s + I_c/m)$ $\quad (2.68)$

$$= (\varepsilon E_c A_c d_c/d_t I_t)\,(I_s + I_c/m)$$

Similarly it may be shown that

$$M_c = \varepsilon E_c A_c d_c I_c/m_c I_t$$
$$M_s = \varepsilon E_c A_c d_c I_s/I_t$$

This method of calculating differential temperature and shrinkage stresses is adopted in CP 117 part 2[6] in which general formulae are quoted for evaluating the stresses due to these effects (see Example 4.5, p.115).

2.8.1 SHRINKAGE

The contraction of concrete which occurs with time, known as shrinkage, has been the subject of study probably as intense as that devoted to creep. Methods of evaluating shrinkage stresses are similar to those already described for creep, and shrinkage strains may be incorporated in the approximate equations of section 2.7.

The value of shrinkage strain to be adopted is dependent on a number of factors summarised in the CEB Recommendations[2] as

$$\varepsilon_{sh} = \varepsilon_c k_b k_e k_p k_t$$

where the notation is

　　ε_{sh} shrinkage strain
　　ε_c environmental coefficient
　　k_b the composition of concrete

k_e the thickness of concrete
k_p percentage of reinforcement
k_t the elapsed time since commencement of loading

k_b and k_t have the same values as for creep (Figure 2.12),
ε_c and k_e are plotted in Figure 2.23,

   k_p is calculated from the expression

   $k_p = 100/(100 + mp)$

where p is the percentage of longitudinal reinforcement in the concrete and m, the modular ratio, is 20.

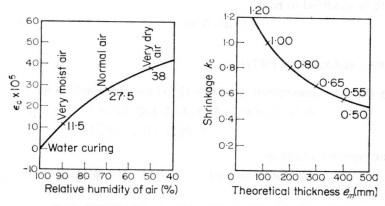

Figure 2.23 Shrinkage coefficients (e_m is defined in Figure 2.12)

2.8.2 EFFECTIVE MODULUS METHOD

Shrinkage strains occur over a period of months so that stresses resulting from them will be modified by creep. The effective concrete modulus should be used in calculating the stresses or the shrinkage strain modified for creep. Various values have been proposed for the modifying factors:

   (a) Effective modulus $E_{ct} = E_c/(1 + 0.52\phi)$

   (b) $E_{ct} = kE_c$ where k is 0.4 to 0.5

   In view of the difficulty of establishing an exact measure of shrinkage strain, the results of using one or other of the modification factors are not likely to produce a significantly more accurate answer. The values of creep coefficient ϕ_t may be taken from Figure 2.12

   The composite section properties are calculated adopting the relevant value of the effective modulus. The stresses due to shrinkage are then obtained from the general differential strain equations.

2.8.3 SATTLER'S APPROXIMATE METHOD

On the assumption that shrinkage strains increase in a manner similar to that for creep we may re-write equation (2.35) to include shrinkage:

$$\lambda \leqslant 0 \cdot 2 \qquad \bar{N} = (N - N_{sh})(1 - e^{-\alpha_s \phi})$$

where

$$N_{sh} = (\varepsilon_{sh}/\phi)E_c A_c$$

Similarly for $\lambda > 0 \cdot 2$ the first term in the top line of equation (2·42) will be modified to read

$$d_t(N_{sh} - N)\phi$$

2.8.4 RELAXATION METHOD

By the same assumption equations (2.51) may be rewritten as

$$\bar{N}_c(1 + \varrho_N \alpha_N \phi) - \bar{M}_c \alpha_N E_c A_c A_s d_t^2 / d_t E_s A_s I_s$$
$$= -\alpha_N \phi[N_c + (\varepsilon_{sh}/\phi)E_c A_c]$$

and equation (2.55) as

$$\bar{N}_c(1 + \varrho_N \alpha_N \phi) = -\alpha_N \phi[N + (\varepsilon_{sh}/\phi)E_c A_c]$$

2.8.5 TEMPERATURE DIFFERENTIAL

The coefficients of expansion of steel and concrete are of comparable order of magnitude so that a freely supported composite beam subjected to a uniform rise in temperature across its full section may undergo little change in stress. Often, however, the slab and beam are not at the same temperature because their differing thermal capacities allow one of them to cool or heat more rapidly than the other. This effect is most marked in bridge and similar structures in which the steel beams are masked from direct solar heating by the slab. Temperature differences of some magnitude can exist in such circumstances.

While in practice the temperature distribution can vary continuously across the beam and slab it is generally accepted that the situation can be represented by assuming that beam and slab are at constant, but different, temperatures. The value of temperature differential to be adopted will depend on the location; for the British Isles ± 10 deg C is proposed in CP 117 part 2. The coefficient of exsion of concrete may be taken as $1 \cdot 1 \times 10^{-5}$/deg C.

2.9 Time-dependent effects in indeterminate composite beams

Time-dependent strain of concrete in a statically determinate beam produces deformation of the beam which is unrestrained by the external support system. This is not the case, however, in statically indeterminate systems. Consider, for example, the two-span composite beam of Figure 2.24; the redundant bending moment at the support may be found by inserting a hinge release at this point and calculating the bending moment required to restore compatibility at the hinge.

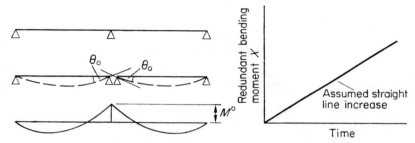

Figure 2.24 Time-dependent effects in indeterminate beams

If the hinge is inserted at time $t = t_o$ and the beam immediately made continuous again, a redundant bending moment M^0 will be required; however, if an interval of time t is allowed to pass before compatibility is restored, the end rotations of the simple beams will have become larger and so the restoring moment M^t will not be the same as M^0. The situation is complicated further by the fact that the composite beam stiffness at time t will be smaller than at time t_0. In the continuous beam, then, M^0 will itself be subjected to creep and so will vary with time. This is equivalent to saying that in addition to the redundant bending moment M^0 a further redundant bending moment X will appear at the support, having zero value at time t_0 and increasing in some way with time.

The problem has been rigorously analysed by Sattler, using Dischinger's equation, but the resulting differential equations are extremely complex and indeed can only be solved for certain simple cases. Sattler therefore proposes a simplified analysis, which takes the following form.

(a) At time t_o determine the bending moment distribution M^0 and thus the sectional forces in the indeterminate beam, $M_s^0, M_c^0, N_s^0, N_c^0$

(b) From these values determine the change in sectional forces in time interval t, $\overline{M}_{st}, \overline{M}_{ct}, \overline{N}_{st}, \overline{N}_{ct}$

(c) With the values $\bar{M}_{st}$ $\bar{M}_{ct}$ $\bar{N}_{st}$ $\bar{N}_{ct}$ find the rotations at each hinge point
(d) Find the redundant bending moments X required at each hinge to restore compatibility. The real value of the redundant bending moments will be less than those calculated in (d) because of creep. Sattler gives a reduction factor ζ by which the calculated values must be multiplied to take account of creep (*Table* 2.5)

Table 2·5 VALUES OF FACTOR ζ

ϕ	ζ
1·0	$0\cdot06\,(A_3/A_4)+0\cdot84+0\cdot003/\alpha_s$
2·0	$0\cdot10\,(A_3/A_4)+0\cdot73+0\cdot004/\alpha_s$
3·0	$0\cdot12\,(A_3/A_4)+0\cdot65+0\cdot005/\alpha_s$

A_3 = steel top flange area, A_4 = steel bottom flange area, $\alpha_s = A_s I_s/A_t I_t$

(e) Using the modified values of the redundant bending moment ζX, corresponding sectional forces M'_s M'_c N'_s N'_c may be calculated
(f) Finally the change in sectional forces M'_s M'_c N'_s N'_c may be found. Because X is itself time dependent, increasing from zero at t_o to a maximum value at t_∞, the change in sectional forces in (d) will be smaller than those occurring under constant bending moment. Sattler proposes a reduction factor of 0·54 to take this into account.

The final condition of the structure is then found by the addition of the stresses calculated for each stage.

Essentially the method proposed by Trost[13] is the same as that outlined above, although the nature of his creep equation allows a direct solution without resort to reduction factors.

2.10 Ultimate load design

In recent years the rational approach of the ultimate load design philosophy has been adopted as the basis of a number of codes of practice. An ultimate load design will permit *failure* to occur only at some specified multiple (load factor) of the working load. The term *failure* in this context may mean any relevant one of a number of possible limit states and it is important to be clear about the limit states which may occur in practice. The situation is complicated, as far as composite beams are concerned, by the multiple nature of the system; concrete slab spanning transversely between steel beams, shear con-

nection, steel beams acting compositely with the slab in the longitudinal direction. Some types of failure which need consideration are

(a) Plastic collapse of the composite beam by the formation of
sufficient hinges
(b) Transverse collapse of the slab by the formation of sufficient
hinge lines
(c) Reduction or complete removal of composite action leading to
beam collapse caused by excessive slip between slab and beam
or complete failure of shear connection
(d) Local shear failure in the slab in regions of high stress around
shear connectors
(e) Excessive deflection

This list by no means contains all possible limit states. However, if
precautions are taken, it is possible to use ultimate load design methods
for a composite beam and slab system, ensuring by over-design that all
but one or two main limit states are suppressed.[7] Research on ultimate
loading of both simply supported and continuous beams has led to the
general conclusion that, provided other modes of failure can be
constrained to occur at a higher load factor, designs may be based
on the assumption of the formation of plastic hinges in the beams.[15, 16]

The continuous beam requires both hogging and sagging plastic
hinges in order that a collapse mechanism may be formed. Experimental work has shown this to be feasible proposition but the sequence
of hinge formation is important; the rotation capacity of sagging
hinges can be curtailed by excessive strain in the concrete and so it is
desirable for hogging hinges, which are not subject to this limitation,
to form first. Considerable progress has been made in formulating
a method for the ultimate load design of continuous composite beams.

2.10.1 ULTIMATE LOAD ANALYSIS

The analysis is based on the assumption that

(a) Initially plane sections remain so after bending
(b) All steel is at yield stress
(c) The concrete stress–strain curve is known (Figure 2.4)
(d) Concrete in tension is ineffective
(e) Slip between beam and slab is negligible

A certain simplification of the real conditions will be apparent here.
The prediction of the ultimate moment capacity of a given beam and
slab combination using the relationships derived below will generally
err on the conservative side because of the neglect of strain hardening
in the steel beam. For a fuller discussion the reader is referred to
Reddy and Hendry.[15]

The ultimate moments of resistance for applied sagging and hogging bending moments may now be determined making use of the known stress–strain relationships and the requirement that the horizontal forces acting on the section must be in equilibrium (Figure 2.25)

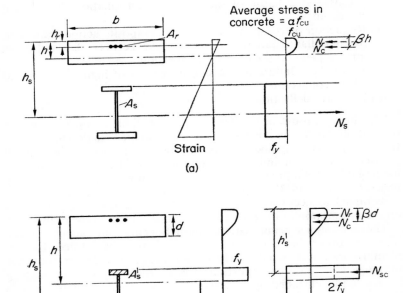

Figure 2.25 Ultimate sagging moment of composite beam. (a) Neutral axis in slab; (b) Neutral axis in beam. A_r = area of reinforcement in concrete slab

Ultimate sagging moment

Neutral axis in slab

$$N_c = \alpha f_{cu} bh$$
$$N_r = A_r f_y$$
$$N_s = A_s f_y$$
$$N_s = N_c + N_r$$
$$A_s f_y = \alpha f_{cu} bh + A_r f_y$$
$$h = f_y(A_s - A_r)/\alpha f_{cu} b$$

For neutral axis to be in slab

$$h < d$$

i.e.

$$(A_s - A_r)f_y/\alpha f_{cu} b < d$$

Ultimate moment of resistance $M_u = A_s f_y(h_s - \beta h) + A_r f_y(\beta h - h_r)$ (2.69)

Neutral axis below slab

$$(A_s - A_r)f_y/\alpha f_{cu} b > d$$

The derivation is simplified if the stress diagram is modified as in Figure 2.25(b) by adding equal and opposite forces $A'_s f_y$ to the area of beam in compression

$$N_c = \alpha f_{cu}bd \qquad N_r = A_r f_y \qquad N_{sc} = 2A'_s f_y$$
$$N_{st} = A_s f_y = N_{sc} + N_r + N_c$$
$$A_s f_y = 2A'_s f_y + A_r f_y + \alpha f_{cu}bd$$
$$A'_s = (A_s f_y - A_r f_y - \alpha f_{cu}bd)/2f_y$$

whence for given steel beam geometry the neutral axis can be found.

$$M_u = A_s f_y(h_s - \beta d) + A_r f_y(\beta d - h_r) - 2A'_s f_y(h'_s - \beta d) \qquad (2.70)$$

Ultimate hogging moment

Neutral axis in beam

The derivation is simplified if the stress diagram is modified as in Figure 2.26 by adding equal and opposite forces $A'_s f_y$ to the area of beam in tension.

$$N_{sc} = A_s f_y$$
$$N_{st} = 2A'_s f_y \qquad N_r = A_r f_y$$
$$N_{sc} = N_{st} + N_r$$
$$A'_s = (A_s f_y - A_r f_y)/2f_y = (A_s - A_r)/2$$

whence for given steel beam geometry the neutral axis can be found.

$$M_u = A_s f_y(h_s - h_r) - 2A'_s f_y(h'_s - h_r) \qquad (2.71)$$

The neutral axis will be in the beam provided $A_s > A_r$. It is extremely unlikely that the tension reinforcement area will exceed that of the steel beam and so the case of the neutral axis in the slab is not considered.

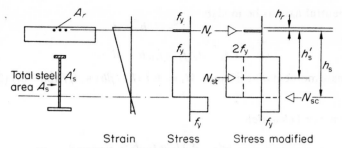

Strain   Stress   Stress modified

Figure 2.26 Ultimate hogging moment of composite beam

Use of equations (2.69) to (2.71) demands knowledge of the parameters α and β for the slab concrete. A simpler method (adopted in
CP 117 part 1)[17] is the use of an equivalent rectangular stress block in
which the average stress is, conservatively, taken as $0\cdot444\,f_{cu}(\alpha=0\cdot444)$
with a resultant at half the height of the rectangle ($\beta=0\cdot5$).

Writing $f_y/0\cdot444\,f_{cu}=k$ and neglecting the reinforcement in the slab:

Neutral axis in slab

$$kA_s/b < d$$
$$h = kA_s/b$$
$$M_u = A_s f_y(2h_s-h)/2 \qquad (2.72)$$

Neutral axis in beam

$$kA_s/b > d$$

The general case will require that the neutral axis depth be found
from the expression for the steel area in compression

$$A_s' = \tfrac{1}{2}(A_s-bd/k) = (kA_s-bd)/2k$$

For beams composed of rectangular elements (Figure 2.27) two
further relationships may be written as follows.

Plastic neutral axis in top flange

Flange area $\qquad A_3 = b_3\times d_3$
$$A_s' < A_3 \qquad (kA_s-bd)/2k < A_3$$
$$h = d+(kA_s-bd)/2kb_3$$
$$M_u = A_s f_y d_t - 2A_s' f_y h/2$$
$$= f_y[A_s d_t - b_3 h(h-d)] \qquad (2.73)$$

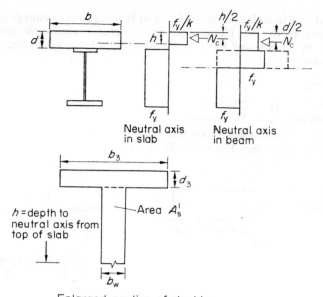

Neutral axis   Neutral axis
in slab   in beam

Enlarged portion of steel beam

Figure 2.27 Simplified ultimate stress diagrams

Plastic neutral axis in beam web

Web thickness b_w

$$A'_s > A_3$$

$$(kA_s - bd)/2k > A_3$$

$$h = d + d_3 + (kA_s - 2kA_3 - bd)/2kb_w$$

$$M_u = A_s f_y d_t - 2A_3 f_y (d + d_3)/2 - 2f_y b_w (h + d_3)(h - d - d_3)/2$$

$$= f_y [A_s d_t - A_3 (d + d_3) - t_w (h + d_3)(h - d - d_3)] \tag{2.74}$$

Note. The above results are valid for the case in which the slab rests directly on the top flange but they may easily be modified to deal with haunched slabs.

EXAMPLE 2.5   ULTIMATE MOMENT OF RESISTANCE OF COMPOSITE UNIVERSAL BEAMS TO CP 117 PART 1

The examples are designed to illustrate the calculation of the ultimate sagging moment of resistance of composite universal beams. The combinations of beam and slab have chosen so that two cases occur;

neutral axis in slab and neutral axis in beam. (Note. Beam properties, obtained from the Handbook on structural steelwork published jointly by BCSA and Constrado, have been rounded to the nearest millimetre.)

Materials

  Steel $f_y = 250$ N/mm²
  Concrete $f_{cu} = 21$ N/mm²

(*a*) Slab 125 mm thick, 1500 mm wide
     Beam 203 mm × 133 mm × 30 kg/m

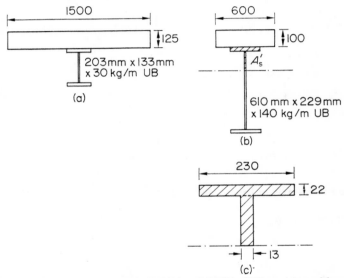

Figure 2.28 Beams for Examples 2.5(a) and 2.5(b). (UB = *universal beam.*)
  Beam data (mm)

A_s	D	b_s	b_w	d_s
3800	207	134	6	10

D is universal beam depth

k $= 250/0.444 \times 21 = 26.8$

$kA_s/b = 26.8 \times 3800/1500 = 68$ mm < 125

i.e. neutral axis lies in the slab and $h = 68$ mm

h_s $(125 + 207/2) = 229$ mm

M_u $[3800 \times 250(2 \times 229 - 68)/2 \times 10^{-6}] = \underline{185 \text{ kN m}}$

(b) Slab 100 mm thick, 600 mm wide
     Beam 610 mm × 229 mm × 140 kg/m.
     Beam data (mm)

A_s	D	b_3	b_w	d_3
17 820	617	230	13	22

$kA_s/b = 26\cdot8 \times 17\ 820/600 = 799 > 100$

i.e. neutral axis lies in beam

Top flange area $A_3 = 230 \times 22 = 5060$ mm²

$A_s' = (26\cdot8 \times 17\ 820 - 600 \times 1000)/2 \times 26\cdot8 = 7790$ mm²

Deduct A_3:   5060

Area of web in compression =   2730 mm²

Depth of web in compression $= 2730/13$ $= 210$ mm

· h $= 210 + 100$ $= 310$ mm

d_t $= (617 + 100)/2$ $= 358$ mm

$A_s d_t$ $= 17\ 820 \times 358$ $= 6\ 400\ 000$

$A_3(d + d_3) = 5060\ (100 + 22)$ $= 617\ 000$

$b_w(h + d_3)\ (h - d - d_3) = 13(310 + 22)\ (310 - 100 - 22) = 811\ 000$

$M_u = 250\ (6\cdot4 - 0\cdot617 - 0\cdot811) \times 10^6$ N mm

   $= 1243$ kN m

2.11 Approximate design of composite plate girders

Initial estimates of the required size of a universal beam which is to
act compositely with a concrete slab can be quickly made using
published tables of beam properties (see for example reference 10,
Chapter 6). The estimation of the dimensions of a composite plate
girder is a more difficult task, even when a designer's accumulated
experience of similar problems can be called upon. However by the
use of simplifying assumptions concerning certain geometrical properties it is possible to arrive at an initial trial section which can fairly
readily be modified into an acceptable working section.

A graphical method is available in Viest, Fountain and Singleton.[18]
It requires, in the initial stage, an estimate of the web size (depth and
thickness) and the ratio of the top and bottom flange areas to the
web area. With this information and the use of the design curves the
initial section can be rapidly checked and, if necessary, improved.

58   FUNDAMENTALS OF COMPOSITE ACTION

When design curves are not available it is possible to proceed entirely from first principles.[19]

Assume first that the web depth d_w can be fixed from span to depth considerations and its thickness b_w from shear and buckling requirements.

$$A_W = d_w \times b_w$$

The equivalent area of concrete A_c/m is also assumed to be known. It will arise from the initial calculations concerned with the design of the slab.

It then remains to determine the areas (A_3 and A_4) of the top and bottom flanges of the girder which are required to resist some imposed bending moment M. To simplify the algebraic expressions, the following further assumptions are made:

  (i) The composite neutral axis lies in the steel beam
  (ii) The flange thicknesses are negligible in comparison with the web depth

The parameter $k = (e + \tfrac{1}{2}d)/d_w$ defines the slab centroid eccentricity in terms of the web depth.

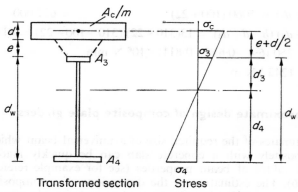

Figure 2.29 Approximate design of composite beam

Referring to Figure 2.29 which defines the geometry of the transformed section (note that σ_c is the equivalent stress on the concrete and is equal to the actual concrete stress multiplied by the modular ratio):

$$d_3 + d_4 = d_w$$

$$d_3 + (e + d/2) = d_3 + k d_w$$

By similar triangles:

$$\frac{\sigma_c}{d_3+e+d/2} = \frac{\sigma_3}{d_3} = \frac{\sigma_4}{d_4}$$

$$d_3 = \left(\frac{\sigma_3}{\sigma_3+\sigma_4}\right)d_w \qquad d_4 = \left(\frac{\sigma_4}{\sigma_3+\sigma_4}\right)d_w$$

$$\sigma_c = \sigma_3 \frac{(d_3+e+d/2)}{d_3}$$

$$= \left(\frac{\sigma_3+\sigma_4}{d_w}\right)\left[\left(\frac{\sigma_3}{\sigma_3+\sigma_4}\right)d_w+kd_w\right]$$

$$= k(\sigma_3+\sigma_4)+\sigma_3 = \sigma_3(1+k)+\sigma_4 \qquad (2.75)$$

For equilibrium

$$\frac{A_c}{m}\sigma_c+A_3\sigma_3+\frac{A_w\sigma_3^2}{2(\sigma_3+\sigma_4)}-\frac{A_w\sigma_4^2}{2(\sigma_3+\sigma_4)}-\frac{A_4}{\sigma_4} = 0 \qquad (2.76)$$

Taking moments about the bottom flange:

$$M = \frac{A_c}{m}\sigma_c\,d_w(1+k)+A_3\sigma_3\,d_w+\frac{A_wd_w}{6}(2\sigma_3-\sigma_4) \qquad (2.77)$$

Combining and rearranging:

$$A_3 = \frac{M}{d_w\,\sigma_3}-\frac{A_w}{6}\left(\frac{2\sigma_3-\sigma_4}{\sigma_3}\right)-\frac{A_ck}{m}\left[2+\frac{\sigma_4}{\sigma_3}+k\left(\frac{\sigma_3+\sigma_4}{\sigma_3}\right)\right]-\frac{A_c}{m}$$
$$(2.78)$$

$$A_4 = \frac{M}{d_w\sigma_4}-\frac{A_w}{6}\left(\frac{2\sigma_4-\sigma_3}{\sigma_4}\right)-\frac{A_ck}{m}\left[\frac{\sigma_3}{\sigma_4}+k\left(\frac{\sigma_3+\sigma_4}{\sigma_4}\right)\right] \qquad (2.79)$$

The first two terms on the right hand side of each expression represent the areas contributed by the steel beam flange and web. The contribution of the concrete is represented by the remaining terms:

$$\text{For } A_3 \qquad \alpha = -\frac{A_c}{m}k\left[2+\frac{\sigma_4}{\sigma_3}+k\left(\frac{\sigma_3+\sigma_4}{\sigma_3}\right)\right]-\frac{A_c}{m} \qquad (2.80)$$

$$\text{For } A_4 \qquad \beta = -\frac{A_c}{m}k\left[\frac{\sigma_3}{\sigma_4}+k\left(\frac{\sigma_3+\sigma_4}{\sigma_4}\right)\right] \qquad (2.81)$$

The total area A_t of the composite section is given by

$$A_t = A_3+A_4+A_w+A_c/m$$
$$= \left(\frac{1}{\sigma_3}+\frac{1}{\sigma_4}\right)\left[\frac{M}{d_w}+(\sigma_3+\sigma_4)\left(\frac{A_w}{6}-k(1+k)\frac{A_c}{m}\right)\right] \qquad (2.82)$$

5*

or, if $\sigma_3 = \sigma_4 = \sigma$,

$$A_t = \left(\frac{2M}{d_w} + \frac{2A_w}{3}\right) - 4k(1+k)\frac{A_c}{m} \qquad (2.83)$$

For given values of M and σ the first two terms on the right hand side represent the area of a steel beam acting alone. The factor $4k(1+k)\,A_c/m$ is the contribution which the concrete makes in reducing the area of a composite beam when compared with the steel beam acting alone.

It should be noted that because the value of k is less than 1 and σ_3 and σ_4 are of the same order, α is generally much larger than β. As might be expected, this fact means that efficient composite beams have much smaller top flanges than bottom flanges.

Initial approximation

Two cases will be investigated — propped and unpropped. Given initial values of the web depth and thickness, and the concrete slab dimensions, the problem is to determine the steel flange areas A_3 and A_4.

The nature of the materials in use will determine the allowable bending stress limits:

Steel in compression f_{sc}
Steel in tension f_{st}
Concrete in
    compression f_{cc}

2.11.1 PROPPED

In this case the whole of the applied bending moment is taken by the composite section. While the bottom flange stress σ_4 may be assumed to be at its maximum value f_{st}, the top flange stress σ_3 and the concrete mean stress σ_c will, in general, not be at their limiting values f_{sc} and f_{cc} simultaneously.

Initially, neglecting k, the top flange stress $= m \times$ (mean concrete stress),

$$\sigma_3 = m\sigma_c$$

and the ruling stresses will be

$$\sigma_4 = f_{st}$$

Criterion	σ_3	σ_c
$f_{sc} > mf_{cc}$	mf_{cc}	f_{cc}
$f_{sc} < mf_{cc}$	f_{sc}	f_{sc}/m

Commonly the concrete dimensions, and thus k, are known initially. Then from equation (2.75),

$$m\sigma_c = \sigma_3 (1+k) + kf_{st}$$

(note σ_c is now the *actual* concrete stress).
Writing $m\sigma_c = mf_{cc}$ (the limiting value),

$$\sigma_3 = \frac{(mf_{cc}-kf_{st})}{1+k} \leqslant f_{sc}$$

and so the following inequalities arise:

Criterion	σ_3	σ_c
$f_{sc} > \dfrac{mf_{cc}-kf_{st}}{1+k}$	$\dfrac{mf_{cc}-kf_{st}}{1+k}$	f_{cc}
$f_{sc} < \dfrac{mf_{cc}-kf_{st}}{1+k}$	f_{sc}	$\dfrac{1}{m}[f_{sc}(1+k)+kf_{st}]$

The values of A_3 and A_4 may then be calculated from equations (2.78 and 2.79). As k is generally small these equations may be simplified without great loss of accuracy to

$$A_3 = \frac{M}{d_w\sigma_3} - \frac{A_w}{6}\frac{(2\sigma_3-\sigma_4)}{\sigma_3} - \frac{A_ck}{m}\left(\frac{2\sigma_3\sigma_4}{\sigma_3}\right) - \frac{A_c}{m} \quad (2.84)$$

$$A_4 = \frac{M}{d_w\sigma_4} - \frac{A_w}{6}\frac{(2\sigma_4-\sigma_3)}{\sigma_4} - \frac{A_ck}{m}\left(\frac{\sigma_3}{\sigma_4}\right) \quad (2.85)$$

2.11.2 UNPROPPED

This case is more complicated because non-composite loading causes stresses in the steel beam alone, and these stresses must be added to those caused by loading on the composite section.

By reasoning similar to that for the propped case it can be shown that for a total bending moment M composed of non-composite load M_g and composite load M_q the following expressions are approximately true:

$$A_3f_{sc} + A_cf_{cc} = \frac{M}{d_w} - \frac{A_w}{6}(2f_{sc}-f_{st}) - kA_cf_{cc} \quad (2.86)$$

$$A_4f_{sc} = \frac{M}{d_w} - \frac{A_w}{6}(2f_{sc}-f_{st}) - kA_cf_{cc} \quad (2.87)$$

$$A_c f_{cc} = \frac{1}{d_w} \left[M - M_g \left(\frac{f_{sc}}{f_{sc} - m f_{cc}} \right) \right] \left[1 + \frac{f_{st}}{f_{sc}} + \frac{6M}{A_w d_w} \right] \quad (2.88)$$

The equations may be solved simultaneously to give the values of A_3, A_4 and A_c which result from the assumption that steel and concrete stresses are at their maximum allowable values.

Notice that the expression for $A_c f_{cc}$ leads to a positive value for

A_c only if $M > M_g f_{sc}/(f_{sc} - m f_{cc})$.

If it leads to a negative value then the concrete stress will not reach its allowable value f_{cc} but will be restricted to

$$\sigma_c = \frac{f_{sc}}{m} \left(\frac{M - M_g}{M} \right) = \frac{f_{sc}}{m} \left(\frac{M_q}{M} \right)$$

In practice A_c will be defined at the commencement of the design of the composite beam, and so the assumption that the allowable stresses f_{sc} and f_{cc} will exist together at working load will not be true. In this case the working values of σ_c and σ_3 may be found from

$$\sigma_c = \frac{f_{sc}}{m} \left(\frac{M_q}{M} \right) \ngtr f_{cc}$$

2.11.3 EXAMPLE OF APPROXIMATE DESIGN METHOD

Material properties

Steel $f_{sc} = f_{st} = 165 \text{ N/mm}^2$
Concrete $f_{cc} = 10 \text{ N/mm}^2$
$m = 15$

Loading

Non-composite dead load $M_g = 3 \times 10^3 \text{ kN m}$

Composite live load $M_q = 7 \times 10^3 \text{ kN m}$

Preliminary considerations of span/depth, shear and buckling lead to the adoption of a web 2000×12 mm. The concrete slab has a thickness d of 160 mm and a width b of 2000 mm.

Steel section alone

$$A_w = (2000 \times 12) \qquad\qquad = 2\cdot4 \times 10^4 \text{ mm}^2$$

Total moment $M = [(3+7) \times 10^3] \qquad = 10^4 \text{ kN m}$

$$A_3 = A_4) = \frac{10^4 \times 10^6}{2\cdot0 \times 1\cdot65 \times 10^5} - \frac{2\cdot4 \times 10^4}{6} \qquad = 2\cdot63 \times 10^4 \text{ mm}^2$$

$$A_s = (2\cdot4 + 2 \times 2\cdot63)10^4 \qquad\qquad = \underline{7\cdot66 \times 10^4 \text{ mm}^2}$$

Note. The assumption of equal areas of top and bottom flange is only justified if the compression flange is restrained from buckling.

Propped

$$k = \frac{160}{2 \times 2000} = 0\cdot04$$

$$\frac{mf_{cc} - kf_{st}}{1+k} = \frac{15 \times 10 - 0\cdot04 \times 165}{1\cdot04} = 141 < 165$$

The working value of σ_3 will be 141 N/mm²

A_3		*Plus*	*Minus*
		(10⁴ mm²)	
$\dfrac{M}{d_w}\sigma_3$	$\dfrac{10^4 \times 10^6}{2\cdot0 \times 1\cdot41 \times 10^5}$	3·54	
$\dfrac{A_w}{6}\left(\dfrac{2\sigma_3 - \sigma_4}{\sigma_3}\right)$	$\dfrac{2\cdot4 \times 10^4}{6}\left(\dfrac{282 - 165}{141}\right)$		0·332
$\dfrac{A_c k}{m}\left(\dfrac{2\sigma_3 + \sigma_4}{\sigma_3}\right)$	$\dfrac{1\cdot6 \times 2\cdot0 \times 4 \times 10^3}{1\cdot5 \times 10}\dfrac{(282+165)}{141}$		0·271
$\dfrac{A_c}{m}$	$\dfrac{1\cdot6 \times 2\cdot0 \times 10^5}{1\cdot5 \times 10}$		2·13
		+3·540	−2·733

$A_3 \quad [= (3\cdot540 - 2\cdot733)10^4] = 0\cdot807 \times 10^4 \text{ mm}^2$

A_4		Plus	Minus
		(10⁴ mm²)	
$\dfrac{M}{d_w \sigma_4}$	$\dfrac{10^4 \times 10^6}{2 \cdot 0 \times 1 \cdot 65 \times 10^5}$	3·03	
$\dfrac{A_w}{6}\left(\dfrac{2\sigma_4 - \sigma_3}{\sigma_4}\right)$	$\dfrac{2 \cdot 4 \times 10^4}{6}\left(\dfrac{330 - 141}{165}\right)$		0·460
$\dfrac{A_c k}{m}\left(\dfrac{\sigma_3}{\sigma_4}\right)$	$\dfrac{1 \cdot 6 \times 2 \cdot 0 \times 4 \times 10^3}{1 \cdot 5 \times 10}\left(\dfrac{141}{165}\right)$		0·073
		+3·030	−0·533

$$A_4 = (3 \cdot 030 - 0 \cdot 533) \times 10^4 = 2 \cdot 497 \times 10^4 \text{ mm}^2$$

Suitable plate sizes would be:

top flange $(410 \times 20) = 0 \cdot 82 \times 10^4$ mm²
bottom flange $(500 \times 50) = 2 \cdot 5 \times 10^4$ mm²

The section properties may then be calculated and stresses evaluated.

Stress		Steel (N/mm²)	Concrete (N/mm²)
σ_1	$\dfrac{1 \times 10^{10}}{61 \cdot 7 \times 10^6 \times 15}$		10·8
σ_2	$\dfrac{1 \times 10^{10}}{72 \cdot 1 \times 10^6 \times 15}$		9·25
σ_3	$\dfrac{1 \cdot 0 \times 10^{10}}{72 \cdot 1 \times 10^6}$	138·6	
σ_4	$\dfrac{1 \times 10^{10}}{60 \cdot 6 \times 10^6}$	165·0	

Unpropped

The propped case was calculated for simplicity using a single value of the modular ratio ($m = 15$) although in practice the section properties for live loading should be calculated using a smaller value of m. Therefore, for the unpropped case, in which only live loading will be imposed on the composite section, a value of $m = 6$ is adopted.

$$\frac{f_{sc}}{m}\frac{M_q}{M}\left(= \frac{165}{6} \times \frac{7}{10}\right) = 19 \cdot 3 \text{ N/mm}^2$$

i.e.

$$\sigma_c = f_{cc} = 10 \text{ N/mm}^2$$

A_3		Plus (10^4 mm^2)	Minus (10^4 mm^2)
$\dfrac{M}{d_w f_{sc}}$	$\dfrac{10^4 \times 10^6}{2 \cdot 0 \times 1 \cdot 65 \times 10^5}$	3·02	
$\dfrac{A_w}{6f_{sc}}(2f_{sc}-f_{st})$	$\dfrac{2 \cdot 4 \times 10^4}{6 \times 165}(165)$		0·400
$A_c \dfrac{f_{cc}}{f_{sc}}(1+k)$	$1 \cdot 6 \times 2 \cdot 0 \times 10^5 \dfrac{10}{165}(1 \cdot 04)$		1·980
		+3·03	−2·380

$$A_3 = (3 \cdot 02 - 2 \cdot 38)10^4 = 0 \cdot 64 \times 10^4 \text{ mm}^2$$

A_4		Plus (10^4 mm^2)	Minus (10^4 mm^2)
$\dfrac{M}{d_w f_{st}}$		3·02	
$\dfrac{A_w}{6f_{st}}(2f_{st}-f_{sc})$			0·400
$\dfrac{kA_o f_{cc}}{f_{st}}$	$\dfrac{0 \cdot 04 \times 1 \cdot 6 \times 2 \times 10^5 \times 10}{165}$		0·078
		+3·02	−0·478

$$A_4 = (3 \cdot 020 - 0 \cdot 478) \times 10^4 = 2 \cdot 542 \times 10^4 \text{ mm}^2$$

Suitable plate sizes would be

Top flange $(430 \times 15) = 0 \cdot 645 \times 10^4$ mm^2
Bottom flange $(520 \times 50) = 2 \cdot 7 \times 10^4$ mm^2

The stresses calculated from these dimensions are given in the following table.

Stress		Steel (N/mm^2)	Concrete (N/mm^2)
σ_1	$\dfrac{7 \times 10^9}{112 \times 10^6 \times 6}$	—	10·4
σ_3	$\dfrac{3 \times 10^9}{2 \cdot 55 \times 10^7} + \dfrac{7 \times 10^9}{135 \times 10^6}$	170	—
σ_4	$\dfrac{3 \times 10^9}{5 \cdot 18 \times 10^7} + \dfrac{7 \times 10^9}{69 \times 100}$	160	—

2.11.4 COMMENTS ON APPROXIMATE DESIGN METHOD

The ratio of dead to live load bending moments in the example is such that there is little to be gained in terms of steel economy from propped construction.

The saving in steel which arises from composite construction is clearly demonstrated, amounting in this example to $\left[\dfrac{(7 \cdot 66 - 5 \cdot 72)}{7 \cdot 66}\right] \times 100 = 25 \cdot 3 \%$ of the weight of a non-composite steel beam.

REFERENCES CHAPTER 2

1. ANDREWS, E. S., *Elementary Principles of Reinforced Concrete Construction*, Scott, Greenwood and Sons, England (1912)
2. *International recommendations for the design and construction of concrete structures*, Cement and Concrete Association, London (1970)
3. YAM, L. C., 'Ultimate load behaviour of composite T-beams having inelastic shear connection', PhD Thesis, Imperial College of Science and Technology, London (Dec. 1966)
4. TIMOSHENKO, S., and GOODIER, J. N., *Theory of elasticity*, 2nd edn, McGraw-Hill, New York, 171–177, (1951)
5. CHAPMAN, J., C. and TERASKIEWICZ, J. S., 'Research on composite construction at Imperial College', *Proc. Conf. on Steel Bridges* British Constructional Steelwork Association, London (1968)
6. *Composite Construction in steel and concrete; beams for bridges*, Code of Practice 117: Part 2, British Standards Institution, London (1967)
7. JOHNSON, R. P., 'Ultimate strength design of sagging moment regions of composite beams', Tech. Rep. S/11, Cambridge University (Aug. 1967)
8. NEVILLE, A. M., *Creep of concrete: plain, reinforced and prestressed*, Elsevier, New York (1970)
9. SATTLER, K., *Theorie der verbundkonstruktionen* (Theory of composite construction), W. Ernst und Sohn, Berlin (1959)
10. NEWMARK, N. M., SIESS, C. P., and VIEST, I. M., 'Tests and analyses of composite beams with incomplete interaction', *Proc. Soc. for exp. Stress Analysis* (1951)
11. HAWRANEK, A., and STEINHARDT, O., *Theorie und berechnung der stahlbrucken* (Theory and design of steel bridges) Springer-Verlag, Berlin (1958)
12. SATTLER, K., 'Composite construction in theory and practice', *Struct. Engr*, **39**, No. 4, (Apr. 1961)
13. TROST, H., 'Zur berechnung von Stahlverbundträgern im gebrauchszustand auf grund neuerer enkenntnisse des viskoelastischen verhaltens des betons' (Design of composite steel girders on the basis of recent investigations of the viscoelastic behaviour of concrete), *Stahlbau* 37 II (Nov. 1968)
14. BUSEMANN, R., 'Kriechberechnung von verbundträgern unter benutzung von zwei kriechfasern' (Calculation of creep effects in loaded composite beams by two creep fibres), *Bauingenieur* 25, **II,** (1950)
15. REDDY, V. M., and HENDRY, A. W., 'Ultimate strength of a composite beam allowing for strain hardening, *Indian Conc. J.* (Sept. 1970)

16. JOHNSON, R. P., VAN DALEN, K., and KEMP, A. R., 'Ultimate strength of continuous composite beams', *Proc. Conf. on Steel Bridges* British Constructional Steelwork Association, London (Sept. 1966)
17. *Composite construction in steel and concrete: simply-supported beams in building,* Code of Practice 117: Part 1, British Standards Institution, London (1965)
18. VIEST, I. M., FOUNTAIN, R. S., and SINGLETON, R. C., *Composite construction in steel and concrete*, McGraw-Hill, New York (1958)
19. FAUCHART, J., and GERBAULT, M., *La construction mixte acier-béton appliquee aux ponts* (Composite construction in steel and concrete for bridges), Office Technique pour L'Utilisation de L'Acier, Paris (1969)

CHAPTER 3

Construction methods

3.1 Introduction

It has already been shown that the full theoretical economy of composite construction can only be achieved if consideration is given to the method by which the structure is to be erected. In common with other types of construction the economics of minimum weight designs are not always as attractive as they might appear on paper, for the reduction in material cost over a simple design may well be overshadowed by the increase in costs of fabrication and erection. Nevertheless, in the right circumstances it is possible to show overall economies because the structural efficiency of composite sections can be considerably improved by adopting a suitable method of construction. Some figures for the reduction in steel weight resulting from the use of propped construction have been quoted earlier. There are other erection techniques aimed at giving a favourable initial prestress to the steel beam or, in continuous beams, an initial compressive stress to the concrete slab. The success of the final structure will be ensured by a judicious choice of materials, design method, structural system and construction method. The importance of the construction method must not be overlooked in the assessment of any scheme.

3.2 Composite action under all types of loading

Because the composite section is more efficient than the steel beam alone it would seem sensible to achieve composite action as early as possible—ideally for all types of loading.

3.2.1 PROPPED CONSTRUCTION

Propped construction, in which the steel beam is supported temporarily until the slab has become composite with it, is an example of one basic type of construction method; the object being to take *all* loading on the composite girder, in contrast with unpropped construction in which the dead load acting before composite action has been achieved is taken on the steel beam alone. 'Propping' need not necessarily mean the provision of temporary supports to the beam in situ: an equally effective method is to place the beam on the ground, cast the slab on it and then erect both beam and slab. This technique has advantages in bridge construction where props may be difficult or impossible to position. The weight limitations of moving a precast beam are also evident. In the construction of the Tay Road Bridge[1], for example, the capacity of the lifting machinery available at the site made it impossible to erect the box beams with a complete concrete deck on them. However, it was possible, while the boxes were continuously supported by the ground, to cast a considerable portion of the slab on them and in this way a useful proportion of the dead load was resisted by composite action. (See also section 7.3.4)

If props are used it is not necessary to provide continuous support to the beam; propping at the quarter span and mid span points will generally be adequate. Indeed the use of just one temporary support can be shown to be beneficial. Consider a simply supported beam of span $4L$ carrying a dead load (steel beam plus concrete slab) g per unit length and a live load q per unit length (Figure 3.1). A temporary support is introduced at midspan point B before construction commences.

At the stage at which the concrete slab has been poured

$$R_{\mathrm{B}} = 2 \cdot 5\, g\, L$$

The bending moments in the steel beam are

$$M_{\mathrm{B}} = -0 \cdot 5\, g\, L^2$$
$$M_{\mathrm{D}} = +0 \cdot 25\, g\, L^2$$

When the concrete slab has hardened the temporary support is removed. Statically the effect of removal is equivalent to applying a force of $-R_{\mathrm{B}}$ at B and leads to bending moments in the composite beam of

$$M_{\mathrm{B}} = 2 \cdot 5\, g\, L\, \frac{4L}{4} = +2 \cdot 5\, g\, L^2$$

$$M_{\mathrm{D}} = \qquad\qquad +1 \cdot 25\, g\, L^2$$

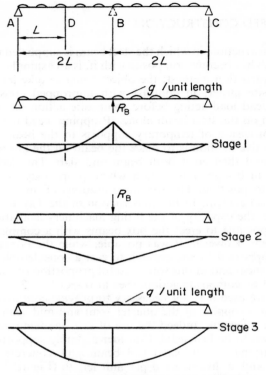

Figure 3.1 Propped construction using central prop

Finally live load gives

$$M_B = +2.0 \; q \; L^2$$
$$M_D = +1.5 \; q \; L^2$$

| | *With prop* | | *Without prop* | |
	Steel	*Composite*	*Steel*	*Composite*
M_B	$-0.5 \, g \, L^2$	$+2.5 \, g \, L^2 + 2 \, q \, L^2$	$+2 \, g \, L^2$	$+2 \, qL^2$
M_D	$+0.25 \, g \, L^2$	$+1.25 \, g \, L^2 + 1.5 \, q \, L^2$	$+1.5 \, g \, L^2$	$+1.5 \, q \, L^2$

demonstrating that a much larger proportion of the bending moment is carried by the composite section when propping is used and the steel beam is correspondingly relieved from bending stress.

Props must be carefully positioned to maintain the beam flange horizontal and rigid enough to resist settlement. The props may be removed as soon as composite action is assured; it can be assumed that this stage is reached when the slab concrete has attained 75 % of its 28 day cube strength. The advantages of propping increase as the ratio of dead to live load increases and for this reason propping may show little economy in building structures.

EXAMPLE 3.1

A simply supported beam spanning	15·0 m
Uniform dead loading	19·0 kN/m
Uniform superimposed loading	37·0 kN/m

The bending moments corresponding to the three cases:

(*a*) without propping
(*b*) with one single central prop
(*c*) continuous propping

are summarised in *Table 3.1*. A distinction is made between long term loading caused by dead load and short term live loading. The relevant value of the modular ratio, *m*, should be used for determining the stresses for dead and live load.

Table 3·1

Case	Bending moment at B (kN m)		
	On steel	On composite section	
		Dead load	Live load
Without prop	+536	—	+1050
With 1 prop	−134	+670	+1050
Continuous props	0	+536	+1050

It will also be necessary, in the case of a single prop, to check the steel beam stresses at D since, before the prop is removed, they will be of opposite sign to those at B.

3.2.2 PHASED CONCRETING

Concrete slabs are commonly constructed in a number of bays rather than in one large pour. By careful choice of the sequence of pour it is possible to achieve composite action early in a critical portion of the beam, using the augmented portion to carry dead load from subsequent pours. An example of phased pouring is shown in Figure 3.2, the sequence of operations being:

1. The steel beam is erected

2. The central one third of the concrete slab is poured and allowed to become composite. With normal curing the time taken to achieve composite action will be about seven days
3. The remaining two thirds of the slab is poured

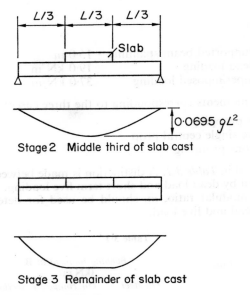

Figure 3.2 Phased concreting

If the weight of slab carried by the beam is g per unit length then, if the whole slab is poured without phasing, the steel beam must carry a bending moment of $0 \cdot 125 \ gL^2$ in addition to its own self-weight bending moment.

By phased concreting, the bending moment on the steel beam (in addition to its own self-weight bending moment) after stage 2 is $0 \cdot 0695 \ gL^2$. In this way $(0 \cdot 0555/0 \cdot 125) \times 100 = 44 \cdot 4\%$ of the slab dead weight bending moment is transferred from the steel beam to the composite section. Whether this is a worthwhile result will depend on the dead-to-live load ratio and the economic practicability of phased casting.

3.3 Prestressing the steel beam

By a number of methods a prestress can be applied to the steel beam in an advantageous sense, i.e. so that its curvature is opposite to that produced by applied loading.[2] Once the prestress has been produced in the beam it may be retained by casting the concrete slab and allowing it to harden, after which the prestressing device can be removed. Alternatively if cables are employed they may be left in place. The required prestress is produced by cambering the steel beam upward over the span using suitable propping or jacking systems or by trussing the beam with prestressing cables. The case of the steel beam prestressed before use in a specialised prestressing rig (the Preflex beam) is considered in Chapter 5.

3.3.1 PRESTRESSING BY JACKING

After the steel beams have been put into position and before any concrete is cast, inner, temporary supports are raised relative to the outer supports. (Some form of tie-down or dead weight loading will be required to hold the beam down at the ends.) The concrete is then placed and allowed to cure, after which the inner supports are removed. At this stage the steel beam has a curvature the reverse of that produced by downward vertical loading. The various stages are shown in Figure 3.3.

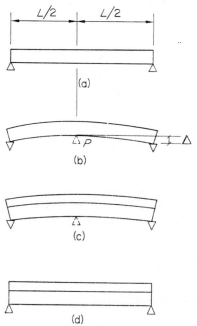

Figure 3.3 Prestressing a composite beam by jacking in situ. (a) Beam erected; (b) Temporary support jacked up; (c) Concrete slab cast; (d) Temporary support removed

The amount by which the beam must be raised can be readily calculated. The limiting factor will be the allowable compressive stress
in the bottom flange of the steel girder and so it will generally be
advantageous to place temporary or permanent transverse bracing
between bottom flanges in position before prestressing, so giving the
greatest allowable compressive stress in the flange.

(a) Single support at midspan

Upward deflection at midspan $\quad \Delta = PL^3/48EI$ $\qquad$ (3.1)

Bending moment at midspan $\quad M = PL/4$ $\qquad$ (3.2)

Compression flange stress $\qquad \sigma_{sc} = My/I$ $\qquad$ (3.3)

Eliminating M and P $\qquad \Delta = \sigma_{sc}L^2/12Ey$ $\qquad$ (3.4)

From (3.1) and (3.4)

$$P = 4\sigma_{sc}I/Ly = 4\sigma_{sc}S_{s4}/L \qquad (3.5)$$

where S_{s4} is the section modulus of the steel beam bottom flange, and
y is its distance from the steel beam centroid.

(b) Two supports each at one third span

$$\Delta = 23PL^3/648EI \qquad (3.6)$$

$$M = PL/3 \qquad (3.7)$$

$$\sigma_{sc} = My/I \qquad (3.8)$$

from which

$$\Delta = 23\sigma_{sc}L^2/216Ey \qquad (3.9)$$

$$P = 3\sigma_{sc}I/Ly = 3\sigma_{sc}S_{s4}/L \qquad (3.10)$$

From these results it is possible to calculate the amount of upward
movement required and the force at each propping point for a given
allowable compressive stress in the steel. As the stress is a temporary
one, a reduced safety factor can be used when calculating the allowable value of compressive stress.

Removal of the temporary supports is equivalent to loading the
composite beam with the jacking forces reversed in direction, or in
other words, with the jacking bending moment reversed. The residual
stress in the steel at this stage (on which the utility of the method
depends) is thus determined by the relative values of the section
moduli for the steel beam acting alone and compositely.

If, for example, the allowable compressive bending stress when the
steel beam is jacked up is σ_{sc} and the section moduli are S_{s3} and S_{s4}

for the steel beam alone and S_{t3} and S_{t4} for the composite beam then the residual stresses after the props have been removed will be

$$\sigma_4 = \sigma_{sc}(1 - S_{s4}/S_{t4})$$
$$\sigma_3 = \sigma_{sc}(S_{s4}/S_{s3} - S_{s4}/S_{t3})$$

In both cases the residual stresses σ_3 and σ_4 will become greater as the ratios *S for steel beam alone/S for composite section* become smaller.

The success of the method depends on achieving a high value of σ_{sc} and a low value of the section modulus ratio. This ratio, for given slab dimensions and selected universal beams, is shown in *Table* 3.2

Table 3·2 COMPARISON OF SECTION MODULUS OF STEEL AND COMPOSITE BEAMS

Beam size		Ratio	Notes
(mm)	(kg/m)	$\dfrac{S_{s3}}{S_{t4}} = \dfrac{S_{s4}}{S_{t4}}$	
152× 89	17·1	0·392	Based on Universal beams
203×133	25·0	0·484	with concrete slabs
251×146	31	0·544	127 mm thick and
304×124	37	0·567	2134 mm wide
304×165	40	0·607	Modular ratio m = 15
356×172	51	0·640	
416×154	74	0·670	
533×209	92	0·716	
545×334	211	0·796	
633×312	238	0·805	

from which it will be seen that relatively small beams with relatively large slabs are best adapted to prestressing by jacking. These conditions will be more generally met in building structures and short span bridges.

EXAMPLE 3.2

A simply supported composite beam spans 10 m carrying a total dead load of 25 kN. The relevant beam properties are shown in Figure 3.4 and *Table* 3.3. Find the maximum allowable live load which it can support:

Table 3·3 PROPERTIES OF COMPOSITE BEAM OF EXAMPLE 3.2 (UNITS mm)

	m	$\dfrac{I}{\times 10^7}$	$\dfrac{S_1}{\times 10^5}$	$\dfrac{S_3}{\times 10^5}$	$\dfrac{S_4}{\times 10^5}$	h
Steel alone	—	4·38	—	2·87	2·87	—
Composite	15	25·0	350	34·7	6·66	107
	7·5	28·1	264	28·4	7·0	80

6*

(a) without prestressing the steel beam

(b) when the steel beam is prestressed by jacking at the third-span points

Material:

    High yield steel f_{sc} = 230 N/mm²

    (but f_{sc} limited to 150 N/mm² for prestressing)

    Concrete f_{cu} = 31 N/mm²

    allowable bending stress   = 10 N/mm²

Without prestress

The calculated steel and concrete stresses under dead load are:

(N/mm²)		*Available for live load* (N/mm²)
σ_1	0	+ 10
σ_3	− 109	− 121
σ_4	+ 109	+ 121

By inspection, the bottom flange stress rules, and so the allowable live load moment, is

$$121 \times 7 \cdot 0 \times 10^5 \times 10^{-6} = 84 \cdot 7 \text{ kN m}$$

giving a maximum live load of 67·9 kN.

With prestress

1. $P = \dfrac{3 \times 150 \times 2 \cdot 87 \times 10^5 \times 10^6}{10} = 12 \cdot 9$ kN

  $\Delta = \dfrac{23 \times 150 \times 10^2 \times 10^6}{216 \times 2 \cdot 1 \times 10^5 \times 153} = 49 \cdot 7$ mm

  $M = \dfrac{12 \cdot 9 \times 10}{3} = 43$ kN m

At the end of this stage $\sigma_3 = -\sigma_4 = +150$ N/mm²

2. After the deck has been cast the props are removed. The stresses due to removal of props are ($m = 15$):

	M/S	(N/mm²)
σ_1	$\dfrac{43 \times 10^6}{3 \cdot 5 \times 10^7}$	− 1·23
σ_3	$\dfrac{43 \times 10^6}{3 \cdot 47 \times 10^6}$	+ 12·4
σ_4	$\dfrac{43 \times 10^6}{6 \cdot 6 \times 10^5}$	+ 64·5

3. The dead load stresses are ($m = 15$):

$$(N/mm^2)$$

σ_1	-0.895
σ_3	$+9.02$
σ_4	$+47.0$

4. The sum of stresses at the end of stage 3 is:

Level	Stress at end of stage 3	Available for LL (N/mm²)
σ_1	-2.13	-7.87
σ_3	$+171$	$+59$
σ_4	-38	-268

5. The allowable live load bending moments based on the stresses available at the end of stage 3 are ($m = 7.5$):

Level	fZ	kN m
σ_1	$7.87 \times 2.64 \times 10^7 \times 10^{-6} = 208$	
σ_3	$59 \times 2.84 \times 10^6 \times 10^{-6} = 167$	
σ_4	$268 \times 7.0 \times 10^5 \times 10^{-6} = 188$	

In this particular case the ruling stress is tension in the top flange, limiting the live load bending moment to 167 kN m or the maximum

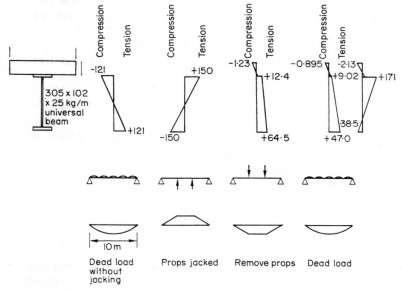

Figure 3.4 Prestressing a composite beam by jacking up intermediate supports

live load to 134 kN, approximately twice that of the case without prestress.

The stress diagrams for each loading case are drawn in Figure 3.4. *Note.* The bending moment due to dead load imposed by the slab while the steel beam is propped should strictly be considered as causing stress in the steel beam. However, because the span is divided into one-third lengths the bending moment will be small and so may be ignored. The situation is analogous to that occurring in propped construction without jacking.

3.3.2 PRESTRESSING THE STEEL BEAM WITH CABLES

Prestressed steel is a form of 'load balancing' in which the beam has a favourable initial stress distribution built into it by the action of high strength cables. By a suitable choice of cable profile a considerable proportion of the imposed bending moment may be balanced out and economy in beam steel obtained at the expense of providing the prestressing system. In certain situations the prestressed steel beam is simpler to install than the cambered version; this is particularly the case in bridges where the propping required for cambering the beam is likely to be difficult or impossible to achieve.

Selection of a suitable cable profile is a compromise between the theoretically most effective profile, in which the eccentricity is varied to produce a prestress bending moment exactly opposed to the service bending moment, and the employment of a less efficient but simpler straight cable of constant eccentricity. The parabolic profile of prestressed concrete tendons is very difficult to achieve with steel beams; the best that can be done is generally a polygonal profile, consisting of straight lines between points of constraint (Figure 3.5).

At changes of direction large forces have to be absorbed and stressing is subject to friction loss. It is thus often preferable to have a straight cable for comparatively short spans of up to say 30 m. The more efficient polygonal cable can be reserved for the longer span beam. Protection of the cables is important; straight cables can be enclosed in the bottom flange of the beam if it is of a suitable hollow shape (see section 5.2). Shaped cables, which must of necessity be exposed, require adequate protection from corrosion.

The stages of construction are

(*a*) Erect steel girders
(*b*) Prestress tendons
(*c*) Cast slab

so that the stress analysis of the system differs only from the conventional by the introduction of bending moment and axial force caused by the prestressing cables. The use of high tensile steel tendons, which

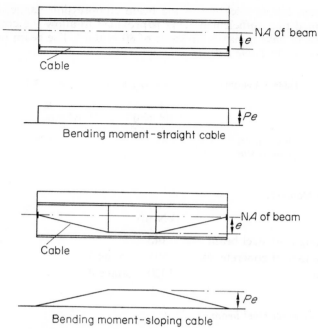

Figure 3.5 Prestressing with cables. P is the force in the cable

have a lower elastic modulus than that of normal structural steels, introduces a slight complication because the area of the tendon must be transformed into an equivalent beam steel area by using the modular ratio

$$\frac{E \text{ for prestressing steel}}{E \text{ for beam steel}} = \frac{E_p}{E_s} = m_p$$

However, where the tendon area is small in relation to the steel beam area, as in the example which follows, the difference can be neglected.

EXAMPLE 3.3

It is known from standard calculations[3] that a high yield steel universal beam 910 mm × 304 mm × 223 kg/m supporting a deck slab 203 mm thick and 1524 mm wide of 31 N/mm² cube strength can be used to carry British Standard Type HA bridge loading on a span of 22·8 m. It is required to calculate the stresses in the same composite beam which is prestressed by cables of constant eccentricity placed in ducts just above the bottom flange and anchored into substantial end plates.

To prevent lateral buckling of the beam compression flange, it will be assumed that sufficient bracing is provided to allow the compressive stress to have a maximum value of 80 N/mm². The beam properties are given in *Table* 3.4.

Table 3.4 PROPERTIES OF COMPOSITE BEAM OF EXAMPLE 3.3

	A ($\times 10^4$ mm)	S_3 ($\times 10^6$ mm)	S_4
Steel alone	2·85	8·24	8·24
Composite		27·6	11·6

Bending Moments

	(kN m)	
Dead weight of steel beam	145	
Dead weight of concrete, etc.	695	— stage 3
Live load	1320	— stage 4

Stage 1 — erect steel beam

$$\sigma_3 = -\sigma_4 = -145 \times 10^6 / 8{\cdot}24 \times 10^6 = -17{\cdot}6 \text{ N/mm}^2$$

Stage 2 — prestress

Let the required prestressing force be P kN at 411 mm eccentricity Uniform axial compressive stress $-P \times 10^3 / 2{\cdot}85 \times 10^4 = -3{\cdot}51 \times 10^{-2}P$ N/mm²
Bending stresses:

$$\sigma_3 = -\sigma_4 = 411P \times 10^3 / 8{\cdot}24 \times 10^6 = 5{\cdot}0 \times 10^{-2}P \text{ N/mm}^2$$

Sum of stresses at the end of stage 2:

$$\sigma_3 = -17{\cdot}6 - 3{\cdot}51 \times 10^{-2}P + 5{\cdot}0 \times 10^{-2}P = (-17{\cdot}6 + 1{\cdot}49 \times 10^{-2}P) \text{ N/mm}^2$$

$$\sigma_4 = +17{\cdot}6 - 3{\cdot}51 \times 10^{-2}P - 5{\cdot}0 \times 10^{-2}P = (+17{\cdot}6 - 8{\cdot}51 \times 10^{-2}P) \text{ N/mm}^2$$

Applying the compressive stress limitation that neither σ_3 nor σ_4 can exceed -80 N/mm², by inspection σ_4 rules.

Solving for P

$$+17\cdot6-8\cdot51\times10^{-2}P = -80$$
$$P = 1150 \text{ kN}$$
$$\sigma_3 = -17\cdot6+17\cdot1 = -0\cdot5 \text{ N/mm}^2$$
$$\sigma_4 = +17\cdot6-97\cdot6 = -80 \text{ N/mm}^2$$

The area of prestressing cable has now to be determined. Because the cable is positioned close to the bottom of the beam it will be subject to relatively large increases in tensile stress when the beam is loaded. For a cable of ultimate strength 1000 N/mm² it is prudent therefore to restrict the initial tensile stress to 500 N/mm², aiming at a maximum value at full load of 700 N/mm².

Area of cable $1150\times10^3/500 = 2\cdot3\times10^3$ mm²

To be strictly accurate the transformed area of the cable should be used to re-compute the steel beam properties.

Transformed area $2\cdot3\times10^3\times1\cdot75\times10^5/2\cdot1\times10^5 = 1\cdot92\times10^3$ mm²
Total steel area $2\cdot85\times10^4+1\cdot92\times10^3 = 3\cdot04\times10^4$ mm²
Shift of neutral axis

$$y = 1\cdot92\times10^3\times411/3\cdot04\times10^4 = 6\cdot3 \text{ mm}$$

Such a small adjustment in the section properties is hardly significant unless extreme accuracy is required. It will be neglected in this example.

The stresses occurring under the remaining loading stages 3 and 4 can now be calculated. They are summarised in Table 3.5 below, with the initial stresses.

Table 3.5

Stresses (N/mm²)

Stage	For this stage			Total at end of stage		
	σ_3	σ_4	σ_p	σ_3	σ_4	σ_p
1	$-17\cdot6$	$+17\cdot6$	—	$-17\cdot6$	$+17\cdot6$	—
2	$+17\cdot1$	$-97\cdot6$	$+500$	$-0\cdot5$	-80	$+500$
3	$-84\cdot5$	$+84\cdot5$	$+77$	$-85\cdot0$	$+4\cdot5$	$+577$
4	-48	$+114$	$+107$	-133	$+119$	$+684$

σ_p = stress in prestressing cable

Comment

By prestressing, the maximum working stress has been reduced from 215 N/mm² to 133 N/mm². The beam material could therefore be mild instead of high yield steel. Alternatively the high yield prestressed beam could be used for a longer span.

3.4 Negative bending of composite sections

The employment of composite beams in situations in which they are subjected to negative (hogging) bending moments produces the difficulty that unless special arrangements are made the whole of the concrete in hogging moment regions will be in tension and thus ineffective. The bending resistance in these regions is reduced to that of the steel beam acting alone and the value of composite action is lost. Cantilevers and continuous beams both produce negative moments but their structural value is such, particularly for longer spans, that efforts should be made to retain composite action in their negative moment areas.

The possible ways of dealing with negative bending are:

    (*a*) deliberately to ensure non-composite action by omitting shear connectors

    (*b*) to place additional steel reinforcement in the deck

    (*c*) to prestress the slab by some suitable method so that under all conditions of loading there is compression in the concrete.

The expedient of adding a concrete slab to the bottom flange of the girder has also been suggested. It seems, however, to be outside the scope of true composite construction in so far as the slab has no other function than to reinforce the girder. If concrete can be used economically, instead of a steel plate, to reinforce a steel girder it would seem that a reinforced concrete girder would be more appropriate than a steel beam and concrete slab.

Several methods are available for inducing an initial compressive force in the slab sufficient to ensure net compression over the full slab depth under all conditions of loading. The use of any particular method will be dependent on the circumstances obtaining at the site but generally it has been found that conventional prestressing using cables is less economic than prestressing by an erection technique. Essentially it is in bridge structures that continuous girders and cantilevers are important; little will be gained by the use of such structural systems in buildings.

3.4.1 NON-COMPOSITE ACTION

In negative moment areas where the steel beam alone will carry the full dead and live loading, a symmetrical section is generally indicated. American practice has in the past tended towards the use of this solution as the least complicated for erection. The analysis of continuous beams is complicated by the existence of an initially unknown length of beam in the negative moment area, which has a reduced stiffness because it is non-composite. An analysis of the continuous composite beam is given by Sherman[4], by which it is possible to calculate the non-composite length. The solution requires finding the roots of cubic and quartic equations and so is not ideally suited to office use. However, it does indicate that it is unwise to neglect the loss of stiffness in negative moment regions by assuming that the composite beam is of constant stiffness over its full span.

It is important to reduce tensile cracking in the slab by deliberately ensuring that there is no composite action between concrete and steel. Natural bond may be prevented by placing building paper or other suitable material between slab and beam. Nominal shear connection is sometimes employed to tie the slab down to the beam.

3.4.2 ADDITIONAL REINFORCEMENT

While the slab will crack in tension in negative moment areas it is still capable of transferring load to steel reinforcement contained by it. Because this additional reinforcement is further from the section neutral axis than the top flange of the beam, a given area of steel will be more effective in the slab than as a beam flange plate. However, because reinforcement only becomes effective after the slab has hardened it makes no contribution to resisting non-composite dead-loading and so, for longer spans where dead loads are high, it may not be practicable.

3.4.3 CABLE PRESTRESSING

A longitudinal compressive force is produced by prestressing tendons either incorporated in the slab or attached to the steel beam. The latter method has already been considered for simple beams.

In regions of negative bending moment, cables are introduced into the slab. The cables may be stressed before or after composite action exists between slab and beam. In the former case the slab is deliberately kept free from the beam by placing the shear connectors in open pockets in the slab, thus ensuring that the prestressing action in the slab is not transmitted to the beam. Once the cables have been stressed the shear connector pockets are grouted up and thenceforth slab and beam act compositely. The advantage of the method is most marked when the deck is composed of precast elements which can be rapidly erected and stressed together. Because the concrete is relatively thin there are difficulties in stressing unless it is prevented from buckling upwards. Some solutions are illustrated in Figure 3.6; these were devised for precast units but they could also be employed with in-situ concrete slabs.

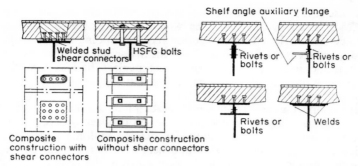

Figure 3.6 Methods of temporary disconnection of slab from beam during prestressing of slab[5]

When prestressing is delayed until the slab is composite with the beam, a different situation occurs. The prestressing force not only produces compression in the slab but, because its line of action is eccentric to the composite neutral axis, it produces bending and compression in the composite beam. The stresses caused by the eccentric force can be calculated by applying the bending and axial force relationships of section 2.6.

The section properties of a prestressed composite beam are calculated in a manner similar to that for its unprestressed counterpart

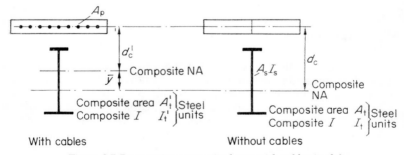

Figure 3.7 Prestressing composite beam with cables in slab

but include an allowance for the transformed area of the prestressing cables. Referring to Figure 3.7:

Equivalent area of prestressing steel $= A_p/m_p$ (3.11)

Total area of prestressed beam $A_t' = A_t + A_p/m_p$ (3.12)

(ignoring area of concrete replaced by prestressing steel).

Vertical shift of neutral axis $\bar{y} = A_p d_c / m_p A_t'$ (3.13)

$$d_c' = d_c - \bar{y} = d_c A_t / A_t'$$ (3.14)

$$I_t' = I_t + A_p d_c'/m_p(1 -$$

$$A_p/mA_t)$$ (3.15)

The factor $\beta = I_s A_s / I_t A_t$ simplifies the following results.

Consider a prestressing force P at the centroid of the concrete slab. It will be distributed as an axial force and bending moment on slab and beam in the following manner (see section 2.6):

Axial force	In cables	P
	In slab	$P(I-\beta)$
	In beam	P
Bending moment	In slab	$P d_t \beta I_c / m I_s$
	In beam	$P d_t \beta$

Prestressing applied to a slab composite with a continuous steel beam leads to a distribution of bending moment which is statically indeterminate. Consider the two span beam of Figure 3.8. The central, composite section is to be prestressed by cables in the concrete at a constant eccentricity d_c from the composite neutral axis. The system may be analysed by the method of virtual work by releasing the beam at B, giving the free bending moment diagram m_o and the reactant diagram $\alpha_1 m_1$. The final bending moment $M_B = m_o + \alpha_1 m_1$. By the virtual work relationship we may write

$$\int M(m_1/EI)\,dx = 0 \quad \text{or} \quad \int (m_0 + \alpha_1 m_1)(m_1/EI)\,dx = 0$$

leading in this particular case to

$$(1/I_1)\int_o^a (x^2/L^2)\alpha_1 dx + (1/I_2)\int_a^L P d_c(x/L)\,dx + (1/I_2)\int_a^L (x^2/L^2)\alpha_1\,dx = 0$$

from which

$$M_B = \left((2b^3[I] - 2b^3 + 3b^2 - 1)/(2b^3[I] - 2b^3 + 2)P d_c\right)$$ (3.16)

where

$$[I] = I_2/I_1$$
$$b = a/L < 1\cdot0$$

From equation (3.16) it will be seen that M_B will always be less than $P d_c$.

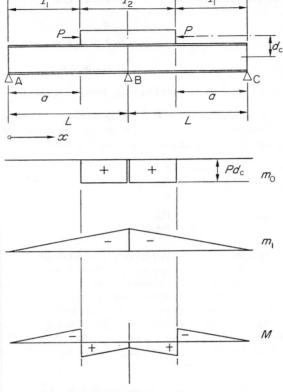

Figure 3.8 Prestressing continuous beam

If the whole beam is made composite before prestressing then $[I] = 1$ and the coefficient in equation (3.16) simplifies to

$$(3b^2 - 1)/2 \qquad\qquad (3.17)$$

Where $b < 1/\sqrt{3}$ equation (3.17) shows that the support bending moment under prestressing will be negative. The cables should therefore be as short as practicable.

EXAMPLE 3.4

A composite beam, continuous over two spans each of 20 m, is to be prestressed in the deck slab by a force sufficient to ensure that tension does not occur in the slab over the central support when the maximum applied negative bending moment at this point is M kN m. Find the prestressing force P required at constant eccentricity d_c from the composite neutral axis

(a) when only 5 m of deck either side of the central support is prestressed
(b) when the full length of the deck is prestressed.

$$I_{composite} = 5I_{steel}$$

(a) Referring to Figure 3.9,

$\int m_o m_1 = [(2 \cdot 5/3)/5I_1](0 \cdot 75 + 4 \times 0 \cdot 875 + 1 \cdot 0)Pd_c = -0 \cdot 875 Pd_c/I_1$

$\int m_1^2 = (7 \cdot 5/3I_1)(4 \times 0 \cdot 375^2 + 0 \cdot 75^2) + (2 \cdot 5/5 \times 3I_1)(0 \cdot 75^2 +$

$\qquad 4 \times 0 \cdot 785^2 + 1 \cdot 0)$

$\qquad = 2 \cdot 8/I_1 + 0 \cdot 77/I_1 = 3 \cdot 57/I_1$

$$\alpha_1 \int m_1^2 + \int m_o m_1 = 0$$

$$\alpha_1 = (0 \cdot 875/3 \cdot 57)Pd_c = 0 \cdot 245 Pd_c$$

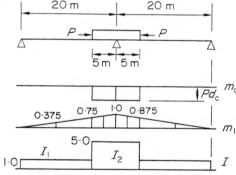

Figure 3.9 Prestressing continuous beam

The bending moment due to prestressing at the central support is given by

$$Pd_c - 0 \cdot 245 Pd_c = 0 \cdot 755 Pd_c \text{ (sagging)}$$

whence

$$P = M/0 \cdot 755 d_c = 2 \cdot 99 M/d_c$$

(b) In this case

$\int m_o m_1 = (-10 Pd_c/3 \times 5I_1)(0 + 4 \times 0 \cdot 5 + 1 \cdot 0)$

$\int m_1^2 = (10/3 \times 5I_1)(0 + 4 \times 0 \cdot 5^2 + 1 \cdot 0)$

$\alpha_1 = 1 \cdot 5 Pd_c$

The bending moment due to prestressing at the central support is given by

$$Pd_c - 1{\cdot}5Pd_c = -0{\cdot}5Pd_c \text{ (hogging)}$$

It is thus not possible to produce a sagging bending moment over the support with a full length cable.

3.4.4 PRESTRESSING BY CAMBERING

If a continuous steel beam is raised above its final level by jacking up over the intermediate supports before the deck is cast on it, and, after composite action has occurred the beam is jacked down to final level, a compressive force will be induced in the slab. The method is similar to that already described for simple beams but in this case the beam *supports* are changed in level to give the required camber. The height of jacking required to produce a given force may readily be calculated. The practical limitation of the method is found when the girder has to be lifted through large distances. However, by the simple expedient shown in Figure 3.10 (below) the difficulty can be overcome[5].

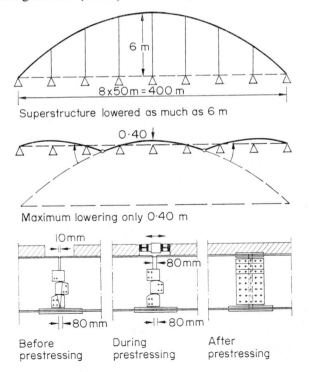

6 m

8 x 50 m = 400 m

Superstructure lowered as much as 6 m

0·40

Maximum lowering only 0·40 m

10mm   80mm

80 mm   80 mm

Before   During   After
prestressing   prestressing   prestressing

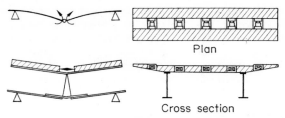

Figure 3.10 Prestressing with temporary link system[5]

As the factor controlling the compressive force induced in the concrete is the initial curvature of the beam, temporary hinges may be inserted at convenient points in the span. The supports may then be raised to achieve the required curvature. As the beam is now effectively a series of much shorter continuous beams, connected by hinges, the amount of jacking is reduced. In the example shown the reduction in lift is from 6·0 m to 0·4 m.

The beam is first erected on level supports with the hinges locked. The hinge position is calculated so that it lies at a point of zero dead-load bending moment; thus when the hinge is unlocked it may be rotated freely. After unlocking the hinges the necessary curvature is produced by jacking the intermediate supports, the deck is cast and finally the composite girder is restored to its level position by lowering the intermediate supports and applying a couple at the hinges by horizontal jacks. The hinges are then permanently fixed by suitable splice plates (Figure 3.11).

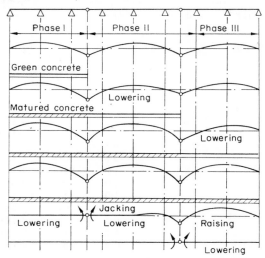

Figure 3.11 Stages in prestressing with temporary link system[5]

Many continuous girder bridges have been erected using jacking techniques. However, it is possible to increase the bending moment in the steel beam at a support by a preloading method which does not demand jacking. Cantilever and suspended span bridges are particularly suited to preloading since the dead weight of the suspended span may be utilised as part of the preload. A typical sequence of operations might be:

(a) Erect all steel girders
(b) Concrete suspended span
(c) Add additional load to suspended span
(d) Concrete cantilever and anchor arm
(e) Remove additional load

The Aboshi Bridge

The details of the bridge given here are taken from an actual structure— the Aboshi Bridge which spans the Ibo River, Japan, and which was erected in 1960[6]. Part of this bridge is a continuous composite plate girder of three equal spans, each of 32 m.

The initial calculations showed that, when the most adverse effects had been taken into account, the support bending moments would produce tension in the slab well above the allowable value.

It was therefore decided to prestress the slab by the cambering technique; a lift of 0·45 m was required. The outer supports were lowered 0·05 m and the inner ones raised 0·40 m, and temporary loads were placed on the girders over the outer supports to prevent

Table 3.6 ABOSHI BRIDGE—CALCULATED STRESSES AT INTERMEDIATE SUPPORT

Load case	Stress (N/mm²)			
	σ_1	σ_2	σ_3	σ_4
1. Dead load	—	—	+116	−86
2. Raising			+103	−76
3. Lowering $t = 0$	−6·8	−3·9	−30	+144
4. Lowering $t = \infty$	−3·6	−2·7	−58	+111
5. Shrinkage	+1·6	+1·5	−16	−23
6. Temperature	±1·0	±0·6	±4	±22
7. Live load	+1·9	+1·1	+8	+41
8. 1+2	—	—	+219	−162
9. 1+2+3	−6·8	−3·9	+189	−18
10. 1+2+4+7	−1·7	−1·6	+169	−110
11. 1+2+4+5+6+7	+0·9	+0·5	+157	−157

Tensile stresses positive

their rising. After concreting (the slab over the inner supports was concreted last) the supports were jacked to their final, level, positions. During the operation the stresses in the concrete and steel were measured. *Table* 3.6 gives the calculated values for all loading cases, i.e. dead and live load, initial and final jacking, creep, shrinkage and a temperature difference of 10 deg C at an intermediate support. The measured stresses after the girder had been lowered were about 80 to 110 per cent of the calculated values and were considered reasonable when allowance was made for the effect of variations in temperature.

REFERENCES CHAPTER 3

1. FAIRHURST, W. A., and BEVERIDGE, 'The superstructure of the Tay Road Bridge', *Struct. Engr.*, **43,** No. 3 (Mar. 1965)
2. HOADLEY, P. G., *The nature of prestressed steel structures* National Research Council, Highway Research Board, Recn 200 (1967)
3. *Simply supported bridges in composite construction*, British Constructional Steelwork Association, London Publication No. BD2 (1970)
4. SHERMAN, J., 'Continuous composite steel and concrete beams', *Trans. Am. Soc. civ. Engrs*, **119,** 810–828 (1954)
5. ROIK, K., 'Methods of prestressing continuous composite girders', *Proc. Conf. on Steel Bridges*, British Constuctional Steelwork Association, London (June 1968)
6. IWAMOTO, K., 'On the continuous composite girder', *Highway Res. Board Bull.* 339, (Bridge deck design and loading studies p. 81) National Academy of Sciences, National Research Council, Washington D.C. (1962)

CHAPTER 4

Shear Connection

4.1 Introduction

Composite action between steel and concrete implies some interconnection between the two materials which will transfer shear between them. In reinforced concrete members the natural bond of concrete to steel is often sufficient to do this, although cases do arise in which additional anchorage is required. The fully encased filler joist also has a large embedded area which is adequate for full shear transfer. However, the situation is quite different with the common type of composite beam in which the concrete slab rests on, or at best encloses, the top flange of the steel beam. It is true that there will initially be shear transfer by bond and friction at the beam–slab contact surface. There is, however, a tendency for the slab to separate vertically from the beam and, should this occur, horizontal shear transfer will cease. A single overload or the fatigue effect of pulsating loading may destroy the natural bond, which once destroyed cannot be reconstituted. The imponderable nature of such shear connection is clearly undesirable; some form of deliberate connection between beam and slab is required with the two objects of transferring horizontal shear and preventing vertical separation. A natural bond will exist in the presence of shear connection but it is neither desirable to count on its existence nor possible in all cases to calculate its value. Thus shear connection must be provided to transfer all the horizontal shear force. It has been pointed out that the paradoxical situation exists that if shear connection is provided it may in fact not come into operation because the natural bond takes all the shear force, and so 'if sufficient shear connectors are provided then they are unnecessary'[1].

The evolution of shear connection devices has been slow and has necessitated a large volume of experimental work on the static and fatigue properties of a wide range of mainly mechanical connectors.

4.2   Mechanical shear connectors

It soon appeared clear to early research workers that some form of
connector fixed to the top flange of the beam and anchored into the
slab was necessary. Caughey and Scott[2] in 1929 proposed using,
amongst other things, projecting bolt ends. Since then a wide variety
of types of mechanical connector has been used in experiment and
practice. To some extent the proliferation of types has been the result
of steel fabricators using sections which came easily to hand, since
initially a purpose-made shear connector was not available.

In any mechanical connection system it is possible to identify parts
which transfer horizontal shear and parts which tie the slab down to
the beam. Generally, horizontal shear resistance is the ruling criterion
of shear connector action and with this in mind mechanical connectors may be classified into three main groups — rigid, flexible and
bond — which are described in more detail in sections 4.4, 4.5 and 4.6.

4.3   Design of mechanical connectors

For a given connector in concrete of a certain strength it is possible to
produce, experimentally, curves relating the shearing load on the
connector to the slip between beam and slab. Typical load–slip curves
for various connectors are plotted in Figure 4.1. The maximum shear
which the connector can sustain may also be determined in this way.
One form of test piece is shown in Figure 4.2; known as the *push-out
test*, it has been adopted in some codes of practice as the standard
specimen for measuring shear connector capacity. Although the specimen does not exactly simulate the actual conditions occurring in a
composite beam or, more particularly, the conditions in the hogging
moment region of a continuous composite beam, it does provide a
useful method of comparing the load–slip and ultimate load capacities
of different types of connector. A double-ended push-out specimen
(Figure 4.3) has been suggested as providing conditions for the shear
connectors closer to those occurring in practice[1].

Tests have also been made on shear connectors embedded in a
concrete slab in tension[3]. The tension in the slab is measured by the
tensile strain in reinforcement, the latter being kept proportional to
the load on the connector. It was concluded that because of wide
variations in the test results further work was required.

If undesirable vertical separation of beam from slab is to be avoided
the connectors must have adequate anchoring properties. The result
of a pull-out test[1] on a stud connector is shown in Figure 4.4. In practice, empirical design rules are used to ensure adequate anchorage
against vertical separation by requiring the connector to have a mini-

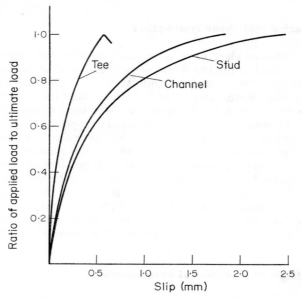

Figure 4.1 Load–slip curves for three different types of shear connector

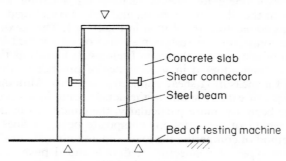

Figure 4.2 Push-out test specimen

mum projection into the concrete. In addition, should there be large forces (loads suspended from the beam for example) tending to cause separation, a special calculation should be made.

The design of shear connectors is not really amenable to theoretical calculation and is best based on the result of push-out or similar tests. Experimental results are available for the more common types of connector both under static and fatigue load. The allowable working load may be calculated from these results either by dividing the ultimate load by a suitable load factor or by finding the load required to produce a specified slip value (say 0·1 mm).

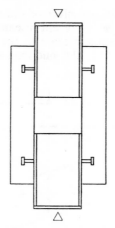

Figure 4.3 Double-ended push-out test specimen

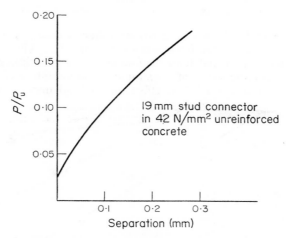

Figure 4.4 Pull-out test on stud connector, P is the force on the connector separating the slab from the beam; P_u is the ultimate push-out capacity of the connector

In addition to the consideration of the load acting on the connector there are important matters which arise from the fact that shear connection between beam and slab is not continuous but concentrated at discrete points. Too great a concentration of shear force may result in cracking of the slab or separation between haunch and slab and so some limit must be placed on local shear stresses in concrete in the neighbourhood of shear connectors. Very large longitudinal distances between connectors may lead in extreme cases to buckling of the slab vertically away from the beam.

The question of the fatigue effect of pulsating load on shear connectors has been the subject of recent research[4]. Considering a simply supported composite beam subjected to the passage of a single concentrated load (Figure 4.5), the end connectors will experience a

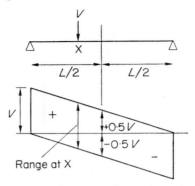

Figure 4.5 Shear force range for single moving load

shear range of V (from V to zero) while the middle connectors will experience a full reversal from $+0.5\ V$ to $-0.5\ V$. Taking account of all the other factors that load the connectors (e.g. shrinkage, temperature change) the ratio of maximum to minimum shear may lie between -1.0 and $+0.5$. The effect on the stud material will be to

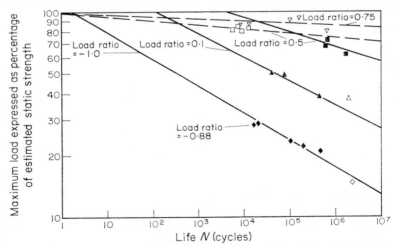

Figure 4.6 Fatigue strengths of stud connectors measured in push-out tests (Building Research Station) in terms of estimated static ultimate strengths.[4] ▽ Load ratio = = 0·75; □ Load ratio = 0·5; △ Load ratio = 0·1; ◊ Load ratio = −1·0 or as marked. Solid symbols represent failures in weld or heat-affected zone. Open symbols represent yield of stud and local crushing of concrete

cause failure to occur at a load less than that obtained in a static push-out test.

The fatigue failure of the connector is quite different from that under static load. The latter occurs generally either as the result of yielding of the connector or crushing of the concrete locally. Fatigue failure is, however, most likely to be caused in the connector-to-beam weld as the result of stress concentration in the weld. Typical curves obtained from dynamic push-out tests on stud connectors are shown in Figure 4.6 and the same results are used to plot the modified Goodman diagram of Figure 4.7. As the result of the research work described by Mainstone[4], design data for shear connectors subjected to pulsating loads have been included in the British Standard Code of Practice 117 Part 2.[5]

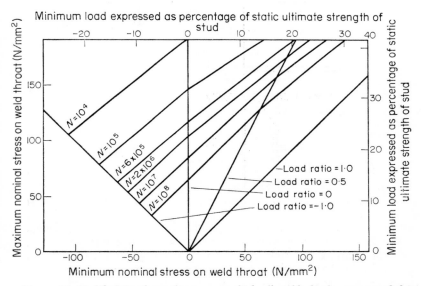

Figure 4.7 Modified Goodman diagram on which allowable loads recommended in Code of Practice 117, Part 2 (BSI) were based

4.4 Rigid Connectors

Because of their inflexible nature, rigid connectors deform very little under load. Slip is small and failure is likely to occur in some other way. In the conditions of concrete containment which the slab concrete undergoes in the region close to a shear connector it is virtually in a state of triaxial stress, and can sustain very high bearing stresses of the order of 350 N/mm². Indeed, failure in service may well occur

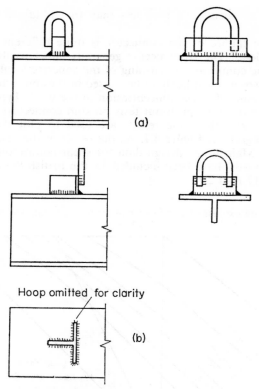

Figure 4.8 Two types of rigid connector. (a) Bar connector; (b) Tee connector

not in concrete bearing but by longitudinal shear in the slab adjacent
to the connector, or by failure of the connector-to-flange weld. Two
common examples of rigid connectors are shown in Figure 4.8, the
tee cutting or bar providing the horizontal shear resistance and the
hoop, vertical tie-down.

Push-out tests on rigid connectors can provide design data, at
least for the capacity of the connector itself, but further calculation
will be required to assess longitudinal shear resistance in the concrete
adjacent to the slab. It is also possible, as described below, to design
rigid connectors from results derived from beam tests. The design
basis used here is that of the German Code of Practice DIN 1078[6].

It is first necessary to assess the allowable bearing stress in the concrete adjacent to the connector face. For a connector of face area a the
the allowable (assumed uniform) bearing stress in the concrete at the
face of the connector is

$$f = f_b(A/a)^{1/3}$$

where f_b is the unrestrained allowable bearing stress in the particular slab concrete. The factor $(A/a)^{1/3}$ gives weight to the spread of bearing stress from the face of the connector into the slab, where A, the distribution area, is defined as

$$A = 2d^2 \quad \text{for unhaunched slabs}$$
$$A = b_o d_o \quad \text{for haunched slabs}$$

In the latter case b_o may not exceed five times the connector width nor d_o five times its height (Figure 4.9).

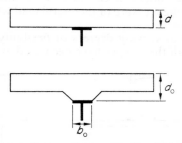

Figure 4.9 Definition of dimensions for rigid connector

The piece of concrete lying between each connector has to develop the force in the connector as a shear stress along its sides and top (Figure 4.10). A situation could occur in which this stress were so high that failure would occur in shear, the implication being that there is a minimum connector spacing below which shear failure will

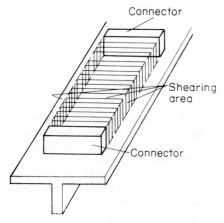

Figure 4.10 Shear on concrete between connectors

occur. Finally, the weld which attaches the connector to the beam
flange must be checked for the shearing and bending effect of the
pressure on its face.

4.5 Flexible Connectors

Unlike the rigid connector which transmits shear between beam and
slab by the bearing of the slab concrete on the connector face, the
flexible connector carries shear primarily in flexure. Slip, which must
be greater when the connector can deform so much more easily, is
commonly taken as the measure of the useful load which the connector
can sustain.

Some connectors of varying degrees of flexibility are illustrated in
Figure 4.11. While all these types have been used at various times it is

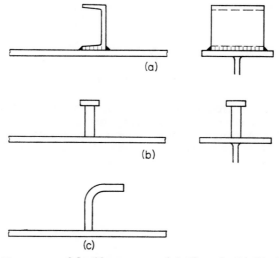

*Figure 4.11 Three types of flexible connector. (a) Channel; (b) Headed stud; (c)
'L' stud*

probably true to say that today the most common of them in use is
the headed stud. The reason for this can be found in the method of
fixing the stud to the beam. Whereas other types of connector must
be conventionally hand welded to the beam flange, the stud is semi-
automatically fixed using a special stud welding gun. The process is
quick and so simple that only semi skilled labour is required.

Details of the stud welding apparatus are shown in Figure 4.12.
The stud forms one electrode of the system, an arc being struck between the bottom end of the stud and the beam flange. A metallic

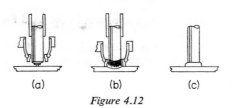

Figure 4.12

flux insert on the stud end ensures a clean weld. After a short delay, during which a pool of molten metal forms under the stud, a spring mechanism in the gun forces the stud into the pool of metal which is retained by a ceramic ferrule placed round the base of the stud. The whole process is completed in a matter of seconds and the gun can then be recharged with a new stud, ready to weld again immediately. The attractions of the system for site welding are particularly clear. It is not, however, suggested that stud connectors are always to be preferred for much must depend on relative labour and material costs.

Design of flexible connectors, whether based on limiting the slip or ultimate load, is only practicable where experimental data are available. *Table* 4.1 gives ultimate capacities for various sizes of stud

Table 4.1 ULTIMATE CAPACITIES OF SHEAR CONNECTORS FOR DIFFERENT CONCRETE STRENGTHS

Type of connector		Connector material	Ultimate capacity (kN)		
			Concrete strength (N/mm²)		
			21	31	41
Headed studs					
Diameter	*Height*				
mm	mm	Minimum			
25	102	yield stress	149	171	194
22	102	386 N/mm²	122	141	160
19	102	Minimum	98	112	127
19	76	tensile stress	84	97	110
16	76	494 N/mm²	71	82	94
13	64		45	52	60
Bars with hoops		BS 4360			
25×38×203 bar		Grade 43	500	750	1000
25×25×203 bar			250	375	500
Channels		BS 4360			
127×64×4·5 kg×152		Grade 43	285	330	375
102×52×3.18 kg×152			262	307	353
76×38×20·4 kg×152			248	293	338
Tees with hoops		BS 4360			
102×76×10×52 high with 13 dia. hoop		Grade 43	253	290	326

connector; from these, by the use of a suitable load factor, a working capacity may be derived.

Local shear failure in the concrete adjacent to the connectors may occur—an example of the method of calculation is given in section 4.9, Example 4.4.

4.6   Bond Connectors

Connectors of the type illustrated in Figure 4.13 are known as bond connectors or anchors, the simplest form of which is a piece of mild steel reinforcement rod butt-welded to the beam flange. They can

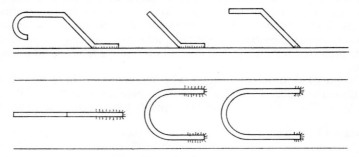

Figure 4.13 Bond connectors

also be combined with a rigid connector to provide tie-down to the slab (composite connector). Horizontal shear transfer between beam and slab takes place by tension in the connector transferred to the concrete by bond.

Various practical considerations determine the use of bond connectors; those illustrated here (Figure 4.14) are abstracted from the German Codes of Practice DIN 1045, DIN 1078[6] and DIN 4239.[7] Adequate bond is ensured by requiring a length of at least thirty dia-

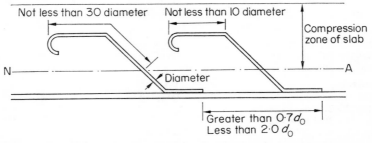

Figure 4.14 DIN rules for bond connectors. For definition of d_0 see Figure 4.9

meters of the bar to be embedded in the compression zone of the slab, of which length at least ten diameters must be horizontal. Minimum and maximum spacings are specified and the depth of the haunch restricted. The connectors can only transfer shear in one direction, because they are ineffective in compression. For this reason, in parts of the beam span subject to reversals of shear, connectors facing in both directions are needed.

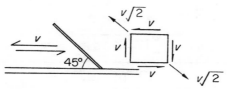

Figure 4.15 Tension on connector at 45° to flange

In the case of pure shear acting on the connector the tensile component acting on the connector placed at 45° (Figure 4.15) is $V(2)^{1/2}$. Thus for a connector of area A and allowable tensile stress f the allowable shear which it can transmit is $fA/(2)^{1/2}$. However, because the angle of inclination of the connector can vary, the allowable shear is increased to fA.

Design then follows the method for other types of connector, with spacing calculated in accordance with the shear force diagram.

4.7 Other types of shear connector

The employment of precast concrete deck units is attractive in many situations, notably temporary structures. Shear connection can be made using conventional connectors in groups, fitting into holes left in the precast units which are subsequently grouted up. However, there are other possible methods, two of which are described below.

4.7.1 HIGH STRENGTH FRICTION GRIP BOLTS

The friction grip (hsfg) bolt is now well established as a steelwork connection, transmitting shearing forces by the friction between the parts it clamps together. In a similar way the hsfg bolt can be used to clamp a concrete slab to a steel beam in the manner shown in Figure 4.16.

The local effect of the tension in the bolt, which could cause crushing failure in the concrete, may be absorbed by spiral reinforcement of the immediate area of concrete around the bolt hole.

Subject to satisfactory bearing of the concrete slab on the steel beam (some form of mortar bedding may be needed) hsfg bolts may be designed as shear connectors with a coefficient of friction of 0·45[8, 9]

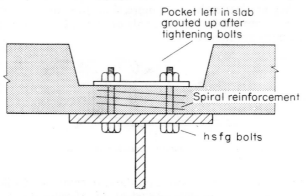

Figure 4.16 High strength friction grip bolt connector

between steel and concrete. The load factor normally adopted for hsfg bolts is 1·7; this may have to be increased for use in bridge decks. (See the second part of Example 4.2.)

4.7.2 GLUING

An apparently simple solution to the shear connection problem is to stick concrete slabs to steel beams with a suitable adhesive. Modern epoxy resin adhesives can provide high strength bond between the two materials and tests have been made on composite beams produced in this way.

The advantages of continuous shear connection rather than the intermittent variety provided by mechanical connectors are obvious. But on the debit side must be mentioned:

    (*a*) Epoxy resin gluing requires conditions of temperature control and cleanliness not usually found on site; it is really a factory operation.

    (*b*) Unlike mechanical connection which penetrates into the concrete and so reinforces it, gluing is restricted to the interface. Failure may then occur in the concrete just above the interface.

It seems at present that more research is required before gluing becomes a practical method of shear connection.

4.8 Longitudinal shear

There is a distinct possibility of longitudinal shear failure in the concrete of composite beams. The relatively wide concrete slab has to receive shear force from the steel beam along a narrow interface; the result may be unacceptably high shear stresses on certain planes in the concrete close to the beam. The shape of concrete haunches is of considerable importance in this respect, since too tall and slender a haunch may be a source of weakness. Codes of practice therefore place limits on haunch dimensions (Figure 4.17).

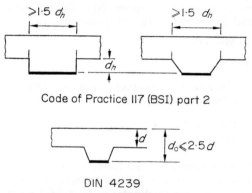

Code of Practice 117 (BSI) part 2

DIN 4239

Figure 4.17 Haunch dimensions

The mean ultimate longitudinal shear strength v_u of a composite beam on any shear plane has two components:

(a) the shear strength of the concrete, which is basically its resistance to tension and is some multiple of the square root of the cube strength $(f_{cu})^{1/2}$
(b) the tensile strength of any reinforcement which may be present in the region

$$v_u = A(f_{cu})^{1/2} + Bpf_y$$

where A and B are constants, p is the total area of transverse reinforcement (intersecting the shear plane) per unit area of shear plane and f_y is the yield stress of the steel

It is essential to ensure that a minimum amount of transverse reinforcement is present at the lower face of the slab as it is in this area that concentrations of stress occur at the shear connectors. Johnson[10] has made a survey of a large number of tests to failure on composite beams and from these, and other reported work, has proposed the following criteria.

(a) All transverse reinforcement (p) contributes to longitudinal shear strength irrespective of its level in the slab and of the magnitude of the negative transverse bending moment.

(b) No account need be taken of longitudinal bending (of either sign) in determining the longitudinal shear strength of a composite beam.

(c) A minimum amount of bottom transverse reinforcement (p_b) is required.

Reduced to specific formulae related to composite beams with normal density concrete slabs, not subjected to fatigue loading or positive transverse bending moments, the requirements are:

$$pf_y \geqslant 1 \cdot 26 v_u - 0 \cdot 28 (f_{cu})^{1/2} \tag{4.1}$$

$$pf_y \geqslant 0 \cdot 552 \tag{4.2}$$

Of the total transverse reinforcement area, not less than half should be placed in the bottom of the slab:

$$p_b f_y \geqslant 0 \cdot 63 v_u - 0 \cdot 14 (f_{cu})^{1/2} \tag{4.3}$$

$$p_b f_y \geqslant 0 \cdot 276 \tag{4.4}$$

The units of these equations are Newtons and millimetres squared and p_b is defined as the area of bottom transverse steel per unit length of slab divided by the slab thickness; $p_b = A_b/d$ (Figure 4.18).

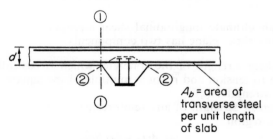

Figure 4.18 Shear planes

For a shear plane which cuts both top and bottom steel the area p will include reinforcement (A_t) already in the slab to resist transverse bending. Provided that this area equals or exceeds half that required by equation (4.2) ($A_t \geqslant p/2$) the only extra reinforcement needed will be at the bottom of the slab, the amount being given by equation (4.4). In the other case the area of top steel must be increased to equal the area of bottom steel given by equation (4.4).

CP 117 parts 1 and 2 (BSI) contains similar but more restrictive clauses which require a minimum percentage of bottom steel but do not allow the steel already present to resist transverse bending to be included. An upper limit on the shear stress in concrete is also given.

CP 117 Part 1

(a) Maximum *ultimate* shear v_u (N/mm²) must not exceed either of:

$$0.625(f_{cu})^{1/2}$$
$$0.234(f_{cu})^{1/2} + 1.0p_b f_y$$

(b) Minimum area of transverse steel $p_b = (v_u/f_y)0.25$

CP 117 Part 2

(a) Maximum *working* shear (N/mm²) must not exceed

$$0.417(f_{cu})^{1/2}$$
$$0.0834(f_{cu})^{1/2} + 0.667p_b f_y \text{ (in sagging moment areas)}$$

(b) Minimum area of transverse steel $p_b = (q/f_y)0.50$

(c) In hogging moment areas the second equation is altered to read

$$(1 - 10\sigma_1/f_{cu})(f_{cu})^{1/2} + 0.667p_b f_y$$

(the first term being ignored if negative) where σ_1 is the maximum longitudinal tensile stress at the top surface of the beam calculated on the assumption that the concrete is uncracked.

4.9 Shear due to differential strain

In addition to the shear transfer required by the bending of a composite beam there is the necessity to absorb shear forces caused by shrinkage and temperature strains. It has been shown in section 2.8 that the shears caused by these effects are confined to the ends of the composite beam and that, for absolutely rigid shear connection, there is, theoretically, an infinitely large interface shear at the beam ends. (In practice, even with very rigid shear connection, the end connectors will yield, throwing load on to connectors further in towards the centre of the beam).

8*

Calculation of end shear due to differential strain may be made directly from equation (2.12) subject to the assignment of a suitable value for the shear connector stiffness μ (see Example 4.5).

However, the calculations involved are rather tedious, and simplifications are adopted in codes of practice.

The German Code DIN 1078 adopts a triangular shear diagram of length equal to the effective width of the beam and area equal to the total shear force on the beam caused by external load (Figure 4.19(a)).

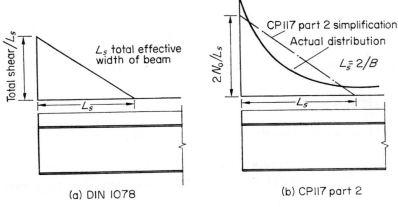

(a) DIN 1078   (b) CP117 part 2

Figure 4.19 Distribution of end shear due to differential strain

The British Code CP 117 part 2 bases its provisions on the exact solutions but simplifies the distribution of shear along the beam from the exponential curve to a straight line (Figure 4.19(b)). The value of shear connector stiffness is fixed (i.e. no account is taken of the actual stiffness) for flexible or rigid connection as shown in *Table* 4.2.

Table 4.2 SHEAR CONNECTOR CONSTANT

Connector	μ (10^{-3} mm²/N)
Stud	2·9
Rigid	1·45

Shear is assumed to fall linearly from a value $2N_o/L_s$ at the end to zero at a distance L_s from the end, where $L_s = 2/B$. (The notation of CP 117 part 2 is different; the transformation to the notation used here is given in Appendix 4.1.) The assumption is that at a distance

$x = 2/B$ from the end of the beam the shear has been sufficiently reduced to be neglected.

We may write the interface shear approximately as $V = N_o B e^{-Bx}$ (see Appendix 4.2). Using this value we have, at a distance $L_s = 2/B$ from the end of the beam,

$$V = N_o B e^{-2} = 0 \cdot 135 N_o B$$

The value of V has thus fallen from $1 \cdot 0 N_o B$ to $0 \cdot 135 N_o B$ and the neglect of any further shear increments is seen to be justified.

EXAMPLES ILLUSTRATING DESIGN OF SHEAR CONNECTORS

The designs are based generally on the requirements of British Standard Code of Practice 117 part 2.

At a point on the beam–slab interface the rate of change of horizontal shear force is given by

$$\frac{dN}{dx} = \frac{d_c A_c}{m I_t} \frac{dM}{dx} = \frac{S_t}{I_t} \times \text{vertical shear force}$$

The quantity S_t/I_t (constant for a beam of given dimensions) is shown as k in the beam section property tables.

Shear connectors are provided primarily to transfer this horizontal shear force between beam and slab. In addition they provide vertical tie-down between the two elements. The spacing of connectors is subject to other constraints:

(a) A minimum centre-to-centre distance determined by the necessity to compact the slab concrete between them.
(b) A maximum centre-to-centre distance above which there is a possibility of local slab deformation.
(c) A minimum projection into the slab to ensure adequate vertical tie-down action.

Conditions of shear transfer between beam and slab also require that a check be made of the shearing stress along certain critical 'shear planes' in the slab, at which shear failure can occur.

The fatigue effect of repeated loading is catered for by using the *range* of shear to design the connectors, rather than the greater of the maximum positive or negative values.

Finally an example is given of the shear produced by shrinkage and temperature difference and the design of special end shear connectors to carry it.

The composite bridge beam section properties are tabulated in *Table* 4.3. Over the end 10 m of the span the vertical shear forces are:

Vertical shear force   (kN)

x (m)	DL		UD		Live load KE		UD+KE–*Range*
	+	−	+	−	+	−	
0	0	134	0	279	0	68	347
5	0	100	9	224	8	55	296
10	0	68	18	176	16	51	26

where DL is the composite dead load, UD the uniformly distributed live load and KE the knife edge live load.

The interface shear may then be calculated, noting that for composite dead load shear (DL) the section properties for $m = 15$ must be used.

x (m)	Interface shear (kN/mm$\times 10^{-2}$)		
	DL	UD+KE	*Total*
0	4·88	15·2	20·1
5	3·62	12·9	16·5
10	2·46	11·1	13·5

Table 4.3 PROPERTIES OF BEAM OF EXAMPLES 4.1–4.5 (UNITS N AND mm)

A_s $\times 10^4$	I_s $\times 10^{10}$	d_t
4·46	2·98	1340

(a)

m	A_c $\times 10^5$	A_c/m $\times 10^4$	A_t $\times 10^4$	I_c $\times 10^8$	I_c/m $\times 10^8$	d_c	I_t $\times 10^{10}$	S_t $\times 10^7$	k $\times 10^4$	E_c $\times 10^4$
7·5	3·61	4·8	9·26	9·88	1·32	648	7·18	3·12	4·35	2·8
15·0	3·61	2·4	6·86	9·88	0·66	876	5·82	2·10	3·62	1·4

(b)

m	$E_c I_c$ $\times 10^{13}$	$E_s I_s$ $\times 10^{15}$	ΣEI $\times 10^{15}$	$\bar{E}I$ $\times 10^{15}$	$E_c A_c$ $\times 10^{10}$	$E_s A_s$ $\times 10^9$	$\bar{E}\bar{A}$ $\times 10^9$	EAd^2_t $\times 10^{15}$
7·5	2·77	6·25	6·28	15·0	1·02	9·35	4·86	8·75
15·0	1·39	6·25	6·27	12·2	0·51	9·35	3·28	5·9

(c)

EXAMPLE 4.1   RIGID CONNECTORS

The connector chosen is a 102 mm$\times$76 mm$\times$9 mm thick tee, 52 mm high, to which is welded a 13 mm diameter bar hoop (overall height to top of bar = 115 mm).

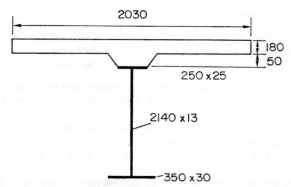

Figure 4.20 Composite beam used in Examples 4.1–4.5. (Dimensions in mm.)

Working capacity

CP 117: Tabulated ultimate capacity = 290 kN (*Table* 4.1)
  Working capacity 290/4 = 73 kN
  DIN 1078: Connector face area a = $100 \times 50 = 5 \cdot 0 \times 10^3$ mm²
Area to which bearing is transmitted $A = 250 \times 230 = 5 \cdot 75 \times 10^4$ mm²
($= b_0 \times d_0$). $A/a = 5 \cdot 75 \times 10^4 / 5 \cdot 0 \times 10^3 > 5 \cdot 0$, i.e. A is restricted to $5a$.

For concrete of cube strength 31 N/mm² the related bearing stress is
$8 \cdot 0$ N/mm², from which $f = 8 \cdot 0 \times (5 \cdot 0)^{1/3} = 13 \cdot 6$ N/mm² ($< 31/2$)
Working capacity $= 13 \cdot 6 \times 5 \cdot 0$ = 68·0 kN

Spacing

Adopt the DIN 1078 working capacity spacing ($= 68 \cdot 0 / 20 \cdot 1 \times 10^{-2}$)
=339 mm. Note. This spacing must be checked to ensure that shear
failure along a shear plane does not occur — see section 4.8.

EXAMPLE 4.2 STUD CONNECTORS

From *Table* 4.1 a suitable stud is 19 mm dia. $\times 102$ mm long
  Ultimate capacity 112 kN
  Because the loading on the beam is standard (HA) bridge loading
the connectors may be designed using the shear range to take account
of fatigue effects.
  Allowable working load ($= 112/4$) = 28 kN (load factor of 4
prescribed by CP 117).
  Using studs in groups of 3:

$$\text{spacing } (= 3 \times 28 / 20 \cdot 1 \times 10^{-2}) = 418 \text{ mm.}$$

Check maximum spacing, least of

Fixed value   = 610 mm
Three times slab thickness (3×180) = 540 mm
Four times stud height (4×100)   = 400 mm

Thus maximum allowable spacing overules the calculating spacing in this case; the connectors may be uniformly spaced along the beam, in threes, at 400 mm centres.

The tie-down requirement of CP 117 requires that the minimum height of a connector (including hoop) shall be the lesser of 100 mm or the depth of the slab minus 25 mm. In this case the connectors must be 100 mm high.

Another possible arrangement of connectors would be in pairs

$$\text{Spacing } 2\times28/20{\cdot}1\times10^{-2} = 279 \text{ mm at end of beam}$$

The spacing could then be increased in steps (Figure 4.21) until the maximum allowable spacing of 400 mm is reached.

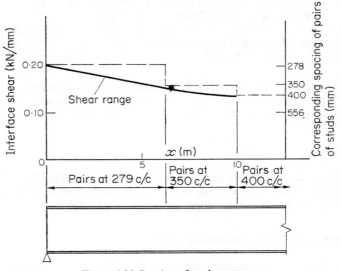

Figure 4.21 Spacing of stud connectors

High strength friction grip bolts

Adopt a 19 mm diameter bolt, for which the proof load (minimum shank tension) is approximately 125 kN.

$$\text{Frictional force for 1 bolt } (= 125\times0{\cdot}45/1{\cdot}7) = 33{\cdot}2 \text{ kN}$$

The bolt spacing can be based on this frictional force value although some assessment of the validity of the load factor of 1·7 may be necessary.

EXAMPLE 4.3 BOND CONNECTORS

Ultimate capacities of bond connectors are not given in CP 117, so that design to that standard would necessitate having available the results of a push-out test.

Adopt the provisions of the German Code.
Allowable tensile stress in mild steel connector: 140 N/mm²
Adopting 16 mm round bar, area 246 mm²
Allowable load on connector ($= 246 \times 140 \times 10^{-3}$) $= 34\cdot4$ kN
Spacing ($= 34\cdot4/20\cdot1 \times 10^{-2}$) $= 176$ mm

The dimension determining the spacing of the connectors is the combined depth of slab and haunch, 230 mm (d_0).

Spacing: $0\cdot7 \times 230 = 161\cdot0$ mm², $2\cdot0 \times 230 = 460$ mm²

so that the calculated spacing is acceptable.
Because of the directional nature of the bond connectors it will be necessary to provide connectors facing in the opposite direction to carry shear in the opposite sense.

EXAMPLE 4.4 SHEAR PLANES

Length of shear planes (see Figure 4.22)

$$L_{1-1} = 180 \text{ mm}$$

$$L_{2-2} (= 210 + 2(70^2 + 50^2)^{1/2}) = 382 \text{ mm}$$

$$L_{3-3} (= 210 + 2 \times 100) = 410 \text{ mm}$$

Areas

$$A_{1-1} (= 180 \times 840) = 1\cdot51 \times 10^5 \text{ mm}^2$$

$$A_{2-2} (= 2 \times 1\cdot51 \times 10^5 + 210 \times 130 + 2 \times 70 \times (180 + 130)/2) = 3\cdot51 \times 10^5 \text{ mm}^2$$

$$A_{3-3} (= 2030 \times 180 + 300 \times 50 - 210 \times 100) = 3\cdot61 \times 10^5 \text{ mm}^2$$

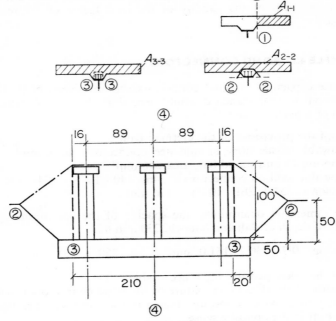

Figure 4.22 Shear planes. (Dimensions in mm.)

Shear plane 1–1

$$Ay/I = 1{\cdot}51 \times 876 \times 10^5/15 \times 5{\cdot}82 \times 10^{10} = 1{\cdot}51 \times 10^{-4} \quad (m = 15)$$
$$= 1{\cdot}51 \times 648 \times 10^5/7{\cdot}5 \times 7{\cdot}18 \times 10^{10} = 1{\cdot}82 \times 10^{-4} \quad (m = 7{\cdot}5)$$

Shear on shear plane $1-1$

DL (kN)	v (kN/mm)	LL (kN)	v (kN/mm)	Total v (kN/mm)
134	$2{\cdot}02 \times 10^{-2}$	347	$6{\cdot}32 \times 10^{-2}$	$8{\cdot}34 \times 10^{-2}$

Allowable shear values (from CP 117 part 2)

(a) $0{\cdot}417 L_s(f_{cu})^{1/2} (= 0{\cdot}417 \times 180 \times (31)^{1/2}) = 0{\cdot}463$ kN/mm
(b) $0{\cdot}0834 L_s(f_{cu})^{1/2} = 0{\cdot}0926$ kN/mm

Thus both criteria are satisfied.

Shear plane 2–2

To be strictly accurate the Ay/I coefficient should be calculated with a value of y measured from the centroid of the excluded area. However, it will be seen from inspection that the rate of shear is approximately $3 \cdot 51/1 \cdot 51 = 2 \cdot 32$ times as great as that on plane 1–1. The allowable shear values will also be increased in the ratio $382/180 = 2 \cdot 12$ and it will be found that the allowable shear stress still exceeds the imposed shear stress.

A similar argument also applies to plane 3–3. The outcome is that no special reinforcement is required. However, in order to provide some restraint, CP 117 stipulates a minimum area of $V/2f_y$ at any shear plane. The worst case will be on a vertical plane on the centre line of the steel beam (4–4, Figure 4.22), where the shear $(20 \cdot 1 \times 10^{-2}/2) = 10 \cdot 1 \times 10^{-2}$ kN/mm. Area of transverse steel required $(10 \cdot 1 \times 10^{-2} \times 10^{-3}/275 \times 2) = 1 \cdot 83 \times 10^{-1}$ mm²/mm. These bars must be spaced at not greater than three times the projection of the shear connectors above them, i.e. $(3 \times 50) = 150$ mm.

Adopt 6 mm bars at 150 mm centres

The bars may be curtailed (subject to suitable anchorage) in accordance with the assumed shear distributions shown in Figure 4.23.

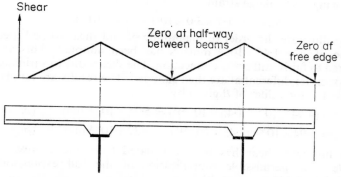

Figure 4.23 Assumed distribution of longitudinal shear for curtailment of transverse reinforcement

EXAMPLE 4.5   SHEAR DUE TO TEMPERATURE DIFFERENTIAL AND SHRINKAGE

The calculation of the interface shear will be made first using the rigorous solution of the partial interaction equations of section 2.5. For shrinkage, which is a long term effect, it is appropriate to use section properties based on the long term modular ratio. Temperature differential on the other hand is a transient phenomenon for which the short term modular ratio should be adopted.

The relevant sums and products of the section properties are given in *Table* 4.3. The parameter B^2 for an, as yet, unspecified value of shear connector stiffness is:

modular ratio m	B^2
7·5	$15 \cdot 0 \times 10^{15}/6 \cdot 28 \times 4 \cdot 86 \times 10^{24} \mu = 4 \cdot 94 \times 10^{-10}/\mu$
15·0	$12 \cdot 2 \times 10^{15}/6 \cdot 27 \times 3 \cdot 28 \times 10^{24} \mu = 5 \cdot 94 \times 10^{-10}/\mu$

On the assumption of unit strain ($\varepsilon = 1$) the values of N_o are:

modular ratio m	N_o (kN/*unit strain*)
7·5	$2 \cdot 8 \times 3 \cdot 61 \times 6 \cdot 48 \times 2 \cdot 99 \times 10^{21}/7 \cdot 18 \times 1 \cdot 34 \times 10^{13} = 2 \cdot 04 \times 10^6$
15·0	$1 \cdot 4 \times 3 \cdot 61 \times 8 \cdot 76 \times 2 \cdot 99 \times 10^{21}/5 \cdot 82 \times 1 \cdot 34 \times 10^{13} = 1 \cdot 7 \times 10^6$

Allowing a temperature differential (slab cooler) of 10 deg C leads to a differential temperature strain of $10 \times 1 \cdot 1 \times 10^{-5} = 1 \cdot 1 \times 10^{-4}$ and

$$N_{o \text{ temp}} (= 2 \cdot 04 \times 1 \cdot 1 \times 10^2) = 2 \cdot 25 \times 10^2 \text{ kN}$$

Allowing a shrinkage strain of $3 \cdot 00 \times 10^{-4}$ gives

$$N_{o \text{ sh}} (= 1 \cdot 7 \times 3 \cdot 0 \times 10^2) = 5 \cdot 1 \times 10^2 \text{ kN}$$

To illustrate the order of magnitude of end shear to be expected a value of $\mu = 1 \cdot 45 \times 10^{-3} \text{ mm}^2/\text{N}$ has been adopted. This value is related to fairly stiff shear connection. The shears due to shrinkage and temperature differential are then summed. The results are tabulated in *Table* 4.4, for values of B given by

$$B (= (4 \cdot 94 \times 10^{-10}/1 \cdot 45 \times 10^{-3})^{\frac{1}{2}}) = 5 \cdot 75 \times 10^{-4} \text{ mm}^{-1} \quad (m = 7 \cdot 5)$$

$$B (= 5 \cdot 94 \times 10^{-10}/1 \cdot 45 \times 10^{-3})^{\frac{1}{2}}) = 6 \cdot 37 \times 10^{-4} \text{ mm}^{-1} \quad (m = 15)$$

The interface shear has been calculated from the expression $V = N_o B e^{-x}$, a permissible simplification of the full expression $V = N_o B \sinh B(1/2 - x)/\cosh B1/2$.

For simply supported beams, CP 117 part 2 requires temperature shear forces to be calculated and shear connectors to be provided to carry these forces. Using the notation of CP 117:

$$\text{Temperature strain} (= 10 \times 1 \cdot 1 \times 10^{-5}) = 1 \cdot 1 \times 10^{-4}$$

$$1/A^1 = 3 \cdot 08 \times 10^6$$

$$A^1 = 3 \cdot 24 \times 10^7$$

$$Q = 2 \cdot 22 \times 10^2 \text{ kN}$$

$$L_s = 3480 \text{ mm}$$

Value of interface shear at end of beam ($= (2\times2\cdot22/3\cdot48)\times10^{-1}$) $= 0\cdot127$ kN/mm The distribution of interface temperature shear on the straight line basis of CP 117 is plotted in Figure 4.24 with, for comparison, the 'exact' distribution.

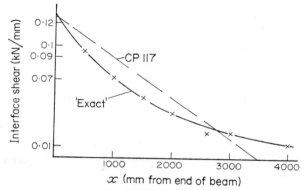

Figure 4.24 Distribution of end shear caused by temperature differential

Table 4.4 VALUES OF INTERFACE SHEAR

x (mm)	Temperature $m = 7\cdot5$				Shrinkage $m = 15$			Total
	Bx	e^{-Bx}	v (kN/mm)		Bx	e^{-Bx}	v (kN/mm)	v (kN/mm)
0	0	1·0	0·1290		0	1·0	0·3250	0·4540
500	0·298	0·742	0·0956		0·319	0·726	0·2360	0·3160
1000	0·575	0·562	0·0725		0·637	0·528	0·1720	0·2445
1500	0·860	0·422	0·0544		0·957	0·383	0·1250	0·1794
2000	1·15	0·317	0·0409		1·27	0·280	0·0912	0·1321
3000	1·72	0·179	0·0231		1·91	0·148	0·0482	0·0613
4000	2·30	0·100	0·0129		2·53	0·080	0·0261	0·0390

APPENDIX 4.1 TRANSFORMATION OF NOTATION FOR SHEAR DUE TO DIFFERENTIAL STRAIN

In CP 117 part 2, use is made of the parameter

$$1/A^1 = d_i^2 + (kI_c/m + I_s)(m/kA_c + 1/A_s)$$

It is required to show that

$$1/A^1 = B^2\Sigma EI \quad \text{where} \quad B^2 = \overline{EI}/\Sigma EI\overline{EA}$$

The factor k modifies the value of m, the instantaneous modular ratio, and may be omitted if we assume m to have a value which is relevant to the differential strain effect being considered.

$$B^2 \Sigma EI = \overline{EI}/\overline{EA} = (\Sigma EI + \overline{EA} d_t^2)/\overline{EA} = d_t^2 + \Sigma EI/\overline{EA}$$
$$\Sigma EI/\overline{EA} = (E_c I_c + E_s I_s)/\{(E_c A_c E_s A_s)/(E_c A_c + E_s A_s)\}$$
$$= (E_c^2 A_c I_c + E_c A_c E_s I_s + E_c I_c E_s A_s + E_s^2 A_s I_s)/E_c A_c E_s A_s$$
$$= E_c I_c/E_s A_s + I_s/A_s + I_c/A_c + E_s I_s/E_c A_c$$
$$= I_c/m A_s + I_s/A_s + I_c/A_c + m I_s/A_c$$
$$= 1/A_s(I_c/m + I_s) + m/A_c(I_c/m + I_s)$$
$$= (I_c/m + I_s)(m/A_c + 1/A_s)$$
$$B^2 \Sigma EI = \; = d_t^2 + (I_c/m + I_s)(m/A_c + 1/A_s)$$

APPENDIX 4.2

$$V = N_o B \sinh B(L/2 - x)/\cosh (BL/2)$$

$$\sinh B (L/2 - x)/\cosh BL/2 = \exp^{\{B(L/2-x)\}} - \exp^{\{-B(L/2-x)\}} \exp^{(BL/2)} + \exp^{(-BL/2)}$$

$\exp^{\{-B(L/2-x)\}}$ and $\exp^{(-BL/2)}$ are small in relation to $\exp^{\{B(L/2-x)\}}$ and $\exp^{(BL/2)}$ because B is large.

Therefore we may write

$$V = N_o B \exp^{\{B(L/2-x)\}} \exp^{(BL/2)} = N_o B \exp^{(-Bx)}$$

REFENCES CHAPTER 4

1. CHAPMAN, J. C., Composite construction in steel and concrete — the behaviour of composite beams', *Struct. Engr*, **42**, No. 4 (Apr. 1964)
2. CAUGHEY, R. A., and SCOTT, W. B., 'A practical method for the design of I-beams haunched in concrete', *Struct. Engr* 47, No. 8 (1929)
3. JOHNSON R., PAUL, *et al.*, 'Stud shear connectors in hogging moment regions of composite beams', *Struct. Engr*, **47**, No. 9 (Sept. 1969)
4. MAINSTONE, R. J., 'Shear connectors in steel-concrete composite beams for bridges and the new CP117 part 2, *Proc. Instn civ. Engrs*, **38**, 83–106 (Sept. 1967)
5. Code of Practice 117 part 2, *Composite construction in steel and concrete: Beams for bridges*, British Standards Institution, London (1967)
6. DIN 1078: 1955, *Specification for the design and development of composite road bridges*, German National Standards Organization
7. DIN 4239: 1956, *Specification for the design and development of composite building structures*, German National Standards Organization
8. SATTLER, K., 'Composite construction in theory and practice', *Struct. Engr*. **39**, No. 4 (Apr. 1961)
9. MARSHALL, W. T., NELSON, H. M., and BANERJEE, H. K. 'An experimental study of the use of high strength friction grip bolts as shear connectors in composite beams', *Struct. Engr*, **49**, No. 4 (Apr. 1971)
10. JOHNSON, R., PAUL., 'Longitudinal shear strength of composite beams', *American Concrete Institute* **67**, pp. 464–6 (June 1970)

CHAPTER 5

Other types of composite construction

5.1 Introduction

Although the composite rolled beam or plate girder is probably the most widely used, there are other types of composite system which merit attention. For special purposes many different composite systems have been proposed; some giving even greater economy in steel weight than the conventional beam and slab, others presenting different advantages. At an early stage in the development of composite construction, novel ideas were put into practice. The Kane system, in use in Canada in 1932, had welded lattice beams which were used to support the shuttering for the concrete floor slab and themselves acted compositely with the slab. In addition, the lattice beams were made continuous, either by passing them through the lattice section columns generally used in the system or, for rolled columns, by using steel rods welded to the top chords of the beams and threaded through holes in the columns. The Kane system thus exhibited many of the economic features of composite construction at an early stage in its development[1].

5.2 Prestressed composite tension flanges

Prestressing a steel girder by cables has already been referred to in Chapter 3; the idea was to produce in the girder a favourable initial stress distribution opposed to that produced in service. The method described in detail here was used for one particular bridge structure[2] but there appears to be no reason why the system should not be of general application, nor indeed why factory methods of production should not be used to produce composite beams of this type.

The girder cross section is shown in Figure 5.1 The 'delta' top girder has advantages in non-composite construction because of the restraint to lateral buckling provided by the corner plates. While compression

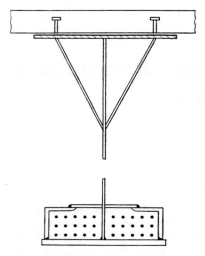

Figure 5.1 Prestressed tension flange

flange restraint is generally not a problem in composite structures (the concrete slab fixes the flange firmly) the delta head is still useful in restraining the girder web. However, the important feature is the hollow bottom flange. Initially the bottom flange is left open. Prestressing wire is threaded through and tensioned, then high strength concrete is vibrated into the flange and steam cured. Finally the flange is converted into a closed box by welding cover plates to it.

The concrete is thus totally enclosed in a steel jacket, free from atmospheric influence. The prestressing wire, too, is entirely protected from corrosion. Because of its enclosure the concrete will not lose moisture and so creep and shrinkage will be minimised. The steel flange is in compression but is adequately restrained from buckling by its concrete filling, the web plate and the prestressing wire.

The advantage of the prestressed flange lies in the much greater stress range available in the steel. If the flange material has an allowable stress in tension or compression of f then the stress range available is from $-f$ to $+f$, a total of $2{\cdot}0f$ compared with $1{\cdot}0f$ for the non-stressed steel.

5.3 Plate girders without top flanges

The unsymmetrical nature of the composite girder with its neutral axis close to the top flange implies low stresses in the top flange steel. Indeed this flange is little more than a means of transferring horizontal shear between beam and slab and could well be omitted altogether if some other method of shear transfer were available.

A section which has been successfully used in practice is composed of an inverted steel 'T' section (rolled or fabricated) to which the concrete slab is fixed by shear connectors welded to each side of the web (Figure 5.2). In fact there is no reason why the bottom flange should not also be replaced by concrete as shown in Figure 5.3

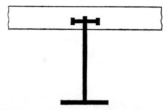

Figure 5.2 Composite plate girder without top flange

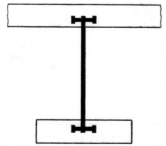

Figure 5.3 Composite plate girder without top or bottom flanges

Tests on a series of hybrid 'T' steel sections of this kind have been reported. Two distinct failure modes were evident, in flexure or in shear originating at the shear connectors. Some tentative design rules aimed at suppressing shear failure were put forward[3].

5.4 Composite lattice girders

The use of open web girders generally leads to economy of steel, though fabrication may significantly affect the final cost of the girder. Two distinct types of open web girder can be distinguished:

9

(a) The very light open web joist used in buildings, often mass produced and being particularly suitable for the passage of pipes and other service ducts

(b) The very much heavier lattice girder or truss used in medium and long span bridges

Research into the behaviour of the light type of joist has shown that, as might be expected, composite action can be achieved between joist and slab if adequate shear connection is provided. In one series of tests[4] a comparison was made between connectors placed either over panel points or midway between them. The conclusion from the test results was that the position of the connectors relative to the panel points was of secondary importance. The load deflection curves of three of the tested joists are shown in Figure 5.4. Joists B1 and B2

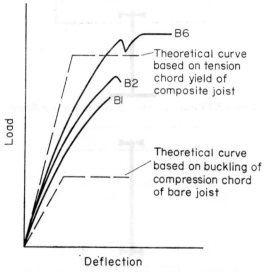

Figure 5.4 Load–deflection curves of three composite open web joists[1]

failed in shear connection; joist B6, having sufficient shear connection, failed in tensile yield of the bottom chord followed by splitting of the concrete. The considerable increase in ultimate load capacity of the composite joist when compared with the steel section alone is also shown in Figure 5.4

The heavier type of bridge truss may also be designed for composite action. A rigorous analysis requires the effect of concrete creep and shrinkage to be taken into account; the analysis is complicated by the indeterminate nature of the slab acting as a continuous beam elastically supported at each panel point. Reference may be made to

Sattler (Chapter 2, reference 9) if the effects of creep and shrinkage are considered important enough to be evaluated.

Because of the eccentricity of the slab relative to the centroid of the steel top chord of the truss, bending will be induced in this chord and distributed at each top chord joint to the other truss members.

5.5 Composite columns

Concrete encasement of steel columns has long been a method of protecting the steel from fire. Just as the concrete encasing filler joist floor beams was long ignored in calculating the beam strength, so concrete column casing until recently was left out of assessment of the steel column strength. Even today there are limits on the assistance which the concrete can be assumed to give. BS 449, ('The use of Structural Steel in Building'), for example, restricts the total axial load on the cased strut to twice that permitted on the uncased steel section. Despite the slowness with which design rules have been officially promulgated, the externally concrete cased steel column is now recognised as a composite section and may be designed as such.

Where a hollow steel section is used as a column a more persuasive argument for composite action with an *internal* filling of concrete can be advanced. The hollow section provides formwork for the concrete and, being itself an efficient compression member, the combination of large diameter high yield steel tube filled with high grade concrete can produce a compact column capable of carrying very large loads. A specific application of such columns (Figure 5.5), occurred at the Almondsbury Interchange where high loads had to be carried on columns with the smallest possible lateral dimensions. The maximum load of 32 000 kN was supported by a composite tube of 1066 mm outside diameter and 44·5 mm thick filled with concrete of cube strength 52 N/mm².[5] On a more modest scale the increase in load carrying capacity of a particular hollow section when filled with concrete is shown in *Table* 5.1. The axial loads given are ultimate values and must be divided by a suitable load factor to give working loads.

The filled hollow section then shows a useful increase in load. It also has an improved fire resistance, although too much should not be made of this aspect as the reduction in the external fire protection required resulting from a concrete infilling is small.

Although many tests on various types of concrete filled hollow sections have been carried out there is not, at the time of writing, a British code of practice for their use. Design must therefore follow one of a number of research reports now available.

A design method for axially loaded columns has been used to compare with a large number of test results reported by several investigators and a reasonable measure of agreement between analysis and

124

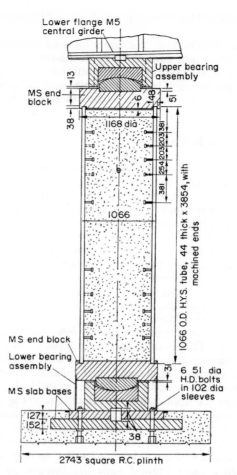

Figure 5.5 Composite column of Almondsbury Interchange[5]. (Dimensions in mm*)*

Table 5.1 COMPARISON OF UNFILLED AND CONCRETE FILLED
HOLLOW STEEL SECTION[7]

Concrete strength (N/mm²)	Ultimate load (kN)
Unfilled	1 000
21	1 220
42	1 440
63	1 690

177·8 × 177·8 × 6·3 mm³ rectangular hollow section
Effective length 3·048 m
Steel yield stress 247 N/mm²

experiment has been shown[6]. For columns with combined axial and bending loads, tables of ultimate loads have been published[7].

It was at one time believed that containment of the infill concrete by the steel casing would lead to much higher ultimate stresses on the infill because it would be in a state of triaxial stress. At the same time the steel casing would be, by the expansion of the concrete, stressed in hoop tension. These effects do occur, but only with short stocky columns.

Axial load design

The basis of the method is the assumption that the ultimate load of the concrete filled column is the sum of the tangent modulus loads of the steel casing and concrete core acting as independent columns.

The general equation for the tangent modulus buckling stress of a column is

$$\sigma_{crit} = \pi^2 E_t / (l/r)^2 \tag{5.1}$$

where E_t is the tangent modulus, and l/r is the slenderness ratio.

The ultimate load of the composite column is thus given by

$$P_u = A_c[\pi^2 E_t / (l/r^2)]_c + A_s[\pi^2 E_t / (l/r)^2]_s \tag{5.2}$$

where the tangent modulus terms refer to the concrete and steel respectively. The ultimate load is defined as soon as the tangent modulus values are known and these will depend on the availability of stress–strain curves from which the tangent modulus may be obtained. A computer program could be written to evaluate the ultimate load of the column, using as data experimentally determined stress–strain relationships. For hand calculation some simplification of actual stress–strain curves will be required.

The design method of Knowles and Park[6] is based on the Hognestad parabolic stress–strain curve for concrete and a parabolic approximation for the variation of tangent modulus load with slenderness ratio for steel. Using these relationships it can be shown that for

$$(l/r)_s < (2\pi^2 E_s / f_y)^{1/2}$$
$$P_u = 2f_c'' A_c q[(q^2+1)^{\frac{1}{2}} - q] + f_y A_s[1 - f_y(l/r)_s^2 / 4\pi^2 E_s] \tag{5.3}$$

and for $(l/r)_s > (2\pi^2 E_s / f_y)^{1/2}$

$$P_u = 2f_c'' A_c q[(q^2+1)^{1/2} - q] + A_s \pi^2 E_s / (l/r)_s^2 \tag{5.4}$$

where f_c'' is the compressive strength of concrete in the column which may be taken as $0.85 \times$ the cylinder strength.

$$q = \pi^2 E_c / 2f_c''(l/r)_c^2$$

E_c is the instantaneous modulus of elasticity of concrete.

It is, however, suggested that for calculating q the cylinder strength of f_c' should be substituted for f_c'', leading to

$$q = \pi^2 E_c / 2 f_c' (l/r)_c^2 \qquad (5.5)$$

EXAMPLE 5.1

The ultimate load for the column of *Table* 5.1 is calculated for comparison using equations (5.3) and (5.4).

Hollow section

$$r = 69{\cdot}8 \text{ mm}, \ A_s = 4{\cdot}34 \times 10^3 \text{ mm}^2$$
$$f_y = 247 \text{ N/mm}^2, \ E_s = 2{\cdot}1 \times 10^5 \text{ N/mm}^2$$
$$(l/r)_s = 3048/69{\cdot}8 = 43{\cdot}8$$
$$(2\pi^2 E_s/f_y)^{1/2} = 129{\cdot}5 \quad \text{(equation (5.3) governs)}$$

Ultimate load on steel alone

$$= 247 \times 4{\cdot}34 \times 10^3 [1 - (247 \times 43{\cdot}8^2)/(4\pi^2 \times 2{\cdot}1 \times 10^5)] \times 10^{-3} \text{ kN}$$
$$(= 1070(1 - 0{\cdot}057)) = \underline{1010 \text{ kN}}$$

Concrete

It is first necessary to fix values for f_c'' and E_c.
Adopting a factor of 0·86 for converting cube strength to cylinder strength:

$$f_c' = (0{\cdot}87 \times 42) = 36{\cdot}5 \text{ N/mm}^2$$
$$E_c' = E_s(u_w)^{\frac{3}{2}}/83 = 1{\cdot}645 \times 10^4 \text{ N/mm}^2$$
$$f_c'' = (36{\cdot}5 \times 0{\cdot}85) = 31{\cdot}1 \text{ N/mm}^2$$

From the geometry of the concrete

$$(l/r)_c = 63{\cdot}8 \qquad A_c = 2{\cdot}73 \times 10^4 \text{ mm}^2$$
$$q = \pi^2 \times 1{\cdot}65 \times 10^4 / 2 \times 36{\cdot}5 \times 63{\cdot}8^2 = 0{\cdot}549$$
$$2q[(q^2 + 1)^{\frac{1}{2}} - q)] = 2 \times 0{\cdot}549(1{\cdot}3^{\frac{1}{2}} - 0{\cdot}549) = 0{\cdot}65$$

Ultimate load on concrete alone

$$(= 31{\cdot}1 \times 2{\cdot}73 \times 10^4 \times 0{\cdot}65 \times 10^{-3})$$
$$= 850 \times 0{\cdot}65 = 552 \text{ kN}$$

Total ultimate load on column is given by

$$P_u = 1010 + 552 = 1562 \text{ kN}$$

Comment

Comparing the results in *Table* 5.1 with this calculation it will be noted that, while the ultimate load on the *steel* column is almost identical by both methods, the tabular result for the filled tube is $(1552-1440) = 122$ kN lower. In part the difference is due to the assumption, in the tabular result, of an ultimate stress in concrete of $0.80 \times cube$ strength.

5.6 Preflex beams

The economy in using high strength steels in rolled sections, which is inherent in their higher allowable bending stresses, may not be realised when deflection is a design criterion. Comparing beams in mild and high yield steels it can be shown that the beam span-to-depth ratio determines whether bending stress or deflection is critical. The

Stage I Precambered beams assembled in pairs

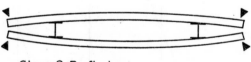

Stage 2 Preflexion

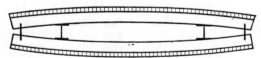

Stage 3 Ends clamped – concrete cast

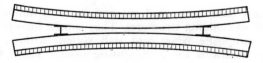

Stage 4 Clamps released

Figure 5.6 Stages in production of a Preflex beam

maximum span-to-depth ratio above which deflection is the ruling design criterion is larger for mild steel beams than for high yield beams. Thus in many circumstances substitution of high yield for mild steel beams will not produce any economy of material.

This fact was appreciated by Lipski and Baes and led to the development of the Preflex beam[8]. Precambered beams of high yield steel are bent in pairs, as shown in Figure 5.6, to a flange stress not exceeding 0·95 of the yield stress. In this condition a high strength concrete flange is cast on the beam tension flange. When the concrete has hardened the preloading is released, the concrete being compressed and some of the initial steel stress being retained. It must be noted that the residual stress in the steel beam is, in fact, in an unfavourable sense since the preflex beam in service is used with the concrete encased flange at the bottom. However, the main point of the system is that with a small increase in beam depth the section stiffness is considerably increased. In addition, the preflexing of the steel beam to such a high stress is an excellent proof test of the material in the beam, and helps to relieve residual stresses present from rolling and other causes[8].

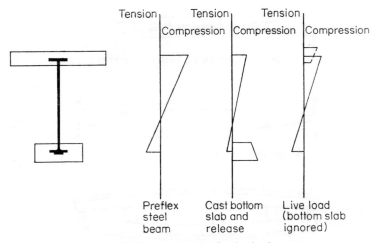

Figure 5.7 Stresses in a Preflex bridge beam

Preflex beams have been used successfully in a number of road bridges as well as building structures. They can be incorporated in a conventional composite beam and slab system without difficulty. Typical stress diagrams for a composite preflex beam bridge are shown in Figure 5.7.

5.7 Composite plates

A steel plate supporting a concrete slab will act compositely by natural bond or by deliberate shear connection. Such composite plates are found, for example, in steel plate bridge decks with a concrete wearing surface or in structures in which ribbed steel sheeting is employed as a form of permanent shuttering. In each case a useful augmentation of strength is available to the deck plate or the concrete slab if composite action exists.

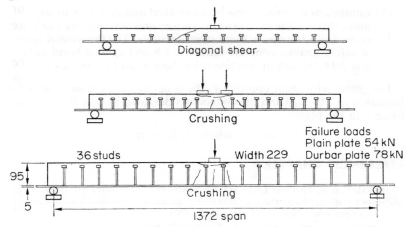

Figure 5.8 Tests on composite plates[9]. (Dimensions in mm.)

Tests have been made on composite plates with shear connectors (Figure 5.8). Under load the connectors are in a zone of concrete in tension, which reduces shear connection efficiency. However, the situation is much improved if plate having an embossed surface or other surface deformity is used. Shear connection must not be too widely spaced as it acts as shear reinforcement to the concrete[9, 10].

As a result of a large number of tests carried out on reinforced concrete slabs supported by some form of profiled steel sheet, it has been established that a useful augmentation of the concrete is provided by the steel sheet through its bond with the concrete leading to a possible reduction in slab reinforcement.

Under increasing load, composite plates show three distinct phases of action. Initially the concrete is fully effective in bending, there is no tensile cracking, and, because the steel plate is very thin, the bond stresses at the steel–concrete interface are very small. As the load increases the concrete eventually cracks in tension. The end of the second phase is marked by sudden loss of bond between steel and concrete. For plates without deliberate shear connection the end of the second phase is marked by total failure. If shear connectors are provided it is possible to increase the load into a third phase at the end of which failure occurs in shear connectors or compressive crushing of the concrete.

It has been suggested[10] that for composite plates without shear connectors the following design criteria should apply:

(*a*) Limitation of compressive bending stress in concrete

$$\sigma_{cc} \leqslant 1 \cdot 67(f_{cu})^{1/2}$$

(*b*) Limitation of bond stress between steel and concrete to 0·05 N/mm^2, treating the section as uncracked and determining the bond stress from the elastic shear stress formula. The theoretical factor of safety against concrete failure is 1·8 and against bond failure it is 2·14, though in practice it has been found to be greater.

For plates with shear connectors the concrete bending stress may be increased to 2·0(f_{cu})$^{\frac{1}{4}}$ and the connectors designed to take the ultimate force P in the plate:

$$P = A_p f_y \qquad \text{where } A_p \text{ is plate area}$$

REFERENCES CHAPTER 5

1. BENTLEY, E. W., Dorman Long Travelling Scholarship *Struct. Engr.* (Feb. 1932)
2. HADLEY, H. M., 'Steel bridge girders with prestressed composite tension flanges', *Civ. Engng.* (Am. Soc. civ. Engrs) (May 1966)
3. TOPRAC, A. A., and EYRE, D. G., 'Composite beams with a hybrid tee steel section', *Proc. Am. Soc. civ. Engrs, Struct. Div.* (Oct. 1967)
4. TIDE, R. H. R., and GALAMBOS, T., ·Composite open web steel joists, *Am. Inst. Steel Constr. Engng. J.*, 7 No. 1 (Jan. 1970)
5. BONDALE, D. S., and CLARK, P. J., Composite construction in the Almondsbury Interchange', *Proc. Conf. on Structural Steelwork*, British Constructional Steelwork Association, London (1966)
6. KNOWLES, R. B., and PARK, R., 'Axial load design for concrete filled steel tubes', *Proc. Am. Soc. civ. Engrs Struct. Div.* (Oct. 1970)
7. *Concrete filled hollow section steel columns design manual*, Comité international pour le développement et l'étude de la construction tubulaire, London (1970)
8. NICHOLAS, R. J., 'Development of the preflexion of beams for bridgeworks', *Proc. Conf. on Steel Bridges*, British Constructional Steelwork Association, London (1968)
9. CHAPMAN, J. C., and TERASZKIEWICZ, J. S., 'Research on composite construction at Imperial College', *Proc. Conf. on Steel Bridges*, British Constructional Steelwork Association, London (1968)
10. CLARKE, J.L., and MORLEY, C. T., 'Steel-concrete composite plates with flexible shear connectors', *Proc. Instn civ. Engrs*, **53**, 557-568 (Dec. 1972)
11. BRYL, S., 'The composite effect of profiled steel plate and concrete in deck slabs', *Acier-Stahl-Steel*, Centre Belgo-Luxemburgeois d'Information de l'Acier, Brussels (Oct. 1967)

CHAPTER 6

Composite construction in buildings

6.1 Introduction

The historical development of composite construction in buildings has been traced in some detail in Chapter 1, showing how the use of concrete as a fire protective casing to iron beams led to the filler joist floor the enhanced strength of which was for a long time neglected. Curiously, although composite construction is the outcome of a structural system widely used in the floors of buildings, it was in bridge works that the first deliberate composite decks were constructed. The innate resistance to change of the building industry, which is described in some detail by Bowley (Chapter 1, reference 1), may have been responsible for the slowness to adopt a new type of structural system in Great Britain; the British Standard Code of Practice (Chapter 2, reference 17) was first published as recently as 1965.

Fire protection regulations (as well as acoustic and other reasons) have made concrete the almost universal material for the construction of floors in buildings. This being so, and the thickness of concrete required for spanning between beams being the same whether the floor is composite or not, there is generally a persuasive argument for the adoption of composite construction because it almost always leads to economy in steelwork. An investigation of the cost of a conventional ten storey office block showed that composite construction could lead to a reduction of approximately 20 % in the weight of the steel framework[1]. Another, less obvious, advantage of composite construction is that the increased stiffness of the beam and slab combination leads to very much reduced deflections. This fact is of importance where the use of high yield steel (which can economically be substituted for mild steel only when it is fully stressed) in a non-composite frame leads to excessive deflection. It must be said, however, that the economics involved is very complicated; for more detailed treatment the reader is referred to the paper by Dorman, Flint and Clark.[2]

Not only the beams may be made composite; columns too can be internally filled, or externally cased in concrete. The filled column is described in Chapter 5; the externally cased column also has an enhanced strength but current design methods do not always take full advantage of the fact.

6.2   Fire Protection

Concrete is still commonly thought of as the most important, or even the only, material for protecting structural steelwork from loss in strength arising from the application of heat. There are, however, good reasons for using concrete where its full strength can be added to that of the element it encases. If this cannot be done the weight penalty of concrete may be too high. When due consideration is given to all the factors involved it will often be found that the floor beams may best be protected by a fireproof suspended ceiling and the columns by some form of hollow lightweight casing. The latter do not require formwork, are of dry construction and are very quickly fixed. Details of typical lightweight fire protection systems have been summarised[3].

6.3   Construction Methods

Many variations of the standard in-situ slab and rolled beam floor have been developed. Speed of construction and use of prefabricated elements are particular advantages which arise from the use of composite construction. The steel beams, unlike conventional concrete reinforcement, can carry their own and other dead loads without needing temporary support from shuttering; by using some form of precast floor unit or permanent shuttering with in situ concrete it is possible to dispense entirely with props between floors. The prefabrication of parts of buildings has reached the stage at which complete floor bays can be lifted into position and bolted up to the columns.

Because of the rough equality of dead and live loads in the average building there is generally little structural advantage to be gained from propped composite construction. Moreover, the props are an embarrassment to finishing operations and so some form of permanent shuttering is indicated. Precast prestressed planks may be used, composite action being produced by the in situ topping concrete laid over the planks and around the shear connectors. Ribbed steel sheet as permanent shuttering has the added advantage that the steel sheet acts compositely with the concrete reducing the amount of slab reinforcement required. With certain restrictions on the thickness of the sheet it is possible to weld stud connectors through it on to the beam flange below.

In multi-storey buildings there is often a need to keep floor depths to a minimum in order to produce the maximum number of storeys in a given height. The use of haunched sections will therefore not generally be advisable. The relatively uneconomic rolled section with a bottom flange plate is sometimes used for minimum depth construction, in high yield steel unless deflection is critical. Because of the trend towards a greater density of such services as air conditioning, facilities for running ducts and pipes under floors are increasingly demanded. In such circumstances castellated beams or open web joists may be used at the expense of some increase in depth.

Continuity in beams and rigid frame action are methods adopted in conventional steel framing to reduce beam weight. With certain complications the same methods may be used for composite beams. The plastic design of rigid frames having composite beams has been carried out and working load deflections have been checked in the elastic range (Chapter 1, reference 11).

6.4   Ribbed sheet steel shuttering

The primary purpose of steel sheet shuttering is to provide a robust, inexpensive permanent shutter which does not require propping. It can be used as a working platform while leaving space beneath it entirely unobstructed. If, in addition, it strengthens the concrete floor slab there is a clear additional economic advantage to be found in its use. We are concerned here with the effect which the steel sheet has on composite beam action; an analysis of the interaction of steel sheet and concrete slab will be found in section 5.7.

The ribbed sheet may be placed so that the ribs run parallel with or at right angles to the steel beams. In the former case, provided the sheet is stopped close to the top flange of the beam, it will not interfere with contact between steel beam and concrete slab and the effect on composite action will be limited to a reduction in slab thickness where the ribs of the sheet penetrate the slab.

There will, however, be occasions when it will be necessary, because of the spans involved, to place the sheet with the ribs running across the steel beams. In such cases there is interference with composite action in the zone of shear connection, leading to a reduction in efficiency. At working load the steel stresses and beam deflections will be greater than those computed using a conventional elastic analysis assuming complete interaction. Additionally, because the shear connectors are embedded in concrete which contains hollow cells caused by the ribs of the sheeting, the ultimate strength of the connectors will be less than that found by a push-out test.

Analysis of the problem, supported by experimental work, has been made by Robinson[4, 5] using the partial interaction equations of section 2.5. The results of this work have been incorporated in a design manual containing simple factors by which the reduction in efficiency of a composite beam with cellular deck may be estimated. The factors are based on an experimentally determined shear connection modulus (for 19 mm diameter studs in 21 N/mm² concrete and a sheeting rib depth of 37 mm) of $1·05 \times 10^5$ N/mm, and an ultimate load capacity for the connectors of about 84 kN.

Further research on composite beams with ribbed steel sheeting supporting slabs of lightweight or normal density concrete has been reported by Fisher[7], from which tentative design recommendations, summarised below, have been put forward.

(*a*) Effective width. Where specifications for determining the effective width include consideration of the slab thickness, the full depth should be used and not the thickness above the ribs.

(*b*) Shear connector capacity. The allowable load on a stud connector is

$$N_{\text{rib}} = 0·5(w/h)(E_{\text{c}-l}/E_{\text{c}-\text{n}})^{\frac{1}{2}}N_{\text{sol}}$$

where N_{rib} is the allowable load on a shear connector in ribbed sheet, N_{sol} is the allowable load on a shear connector in solid concrete of normal density, $E_{\text{c}-l}$ is the modulus of elasticity of lightweight concrete, $E_{\text{c}-\text{n}}$ is the modulus of elasticity of normal density concrete and w and h are rib dimensions (Figure 6.1)

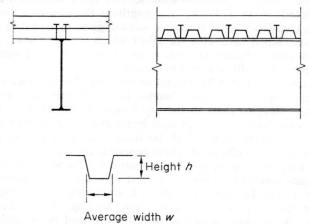

Average width *w*

Figure 6.1 Composite beam slab supported on ribbed steel sheeting

(*c*) Where the height *h* of the ribs is less than 37 mm and the bottom of the compression zone in the slab does not extend below the rib, the slab may be treated as having its full thickness.

(*d*) In all other cases the section properties are affected by the ribs. Concrete below the top of the ribs is not considered effective and so the section is reduced by the loss of concrete area in compression. In both elastic and ultimate load analysis the depth to the bottom of the compressive stress block should be determined. If this value exceeds the depth of concrete over the top of the rib then the composite beam should be analysed as if only concrete over the top of the ribs is effective (see Example 6.6).

(*e*) Shear connectors should be as long as possible, extending at least 37 mm above the top of the ribs.

6.5 Prestressed steel in buildings

Preflex beams are mentioned elsewhere (see section 5.6). Their use in buildings is attractive where there are limitations on the beam depth.

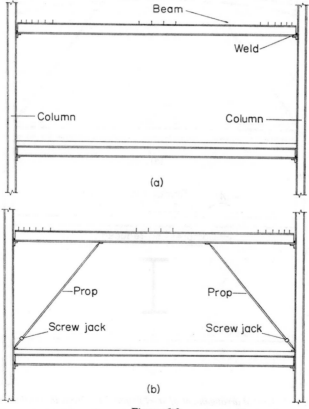

Figure 6.2

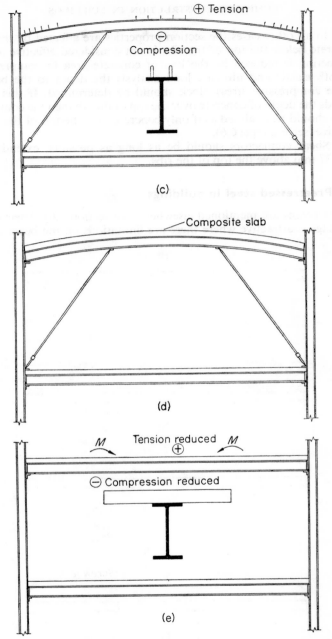

Figure 6.2 (a) General arrangement of steel frame; (b) Props in position; (c) Props extended; (d) Slab cast; (e) Props removed after slab has hardened

It is also possible by a jacking method in situ to produce an upward
camber in the bare steel beams of steel frames in a manner similar
to that described in section 3.3. The system (patented —Chapter 1,
reference 12 Wilenko) is shown diagrammatically in Figure 6.2.
A simply supported steel beam is restrained from upward movement

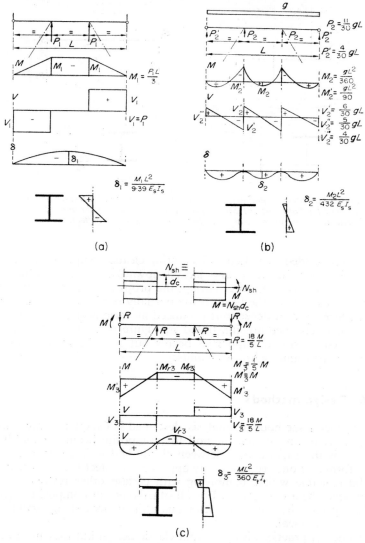

Figure 6.3

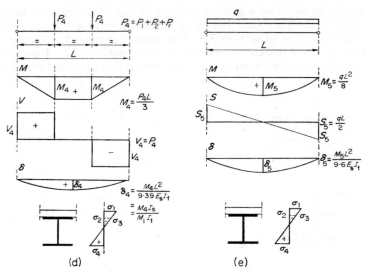

Figure 6.3 (a) Stresses and deflections due to propping; (b) Stresses and deflections due to dead load of slab; (c) Stresses and deflections due to shrinkage; (d) Stresses and deflections due to removal of props; (e) Stresses and deflections caused by live load

by being welded or bolted to its seating cleats. Temporary props are placed under the beam and jacked upwards, following which the concrete slab is cast. After curing, the props are released, the initial upward camber of the beam being partially retained by the concrete slab. Stresses and deflections produced at various stages in the construction sequence are illustrated in Figure 6.3. The system is not restricted to simply supported beams but may be applied also to continuous frames with rigid joints.

6.6   Design methods

The general methods of analysis described in Chapter 2 must form the basis of any design method for composite beams in buildings. There is the usual choice between elastic and ultimate load design; the former is only economically applicable to steel beams encased in solid concrete which, complying with certain other restrictions, do not need shear connectors. For all other types of composite section, which do require shear connectors, an ultimate load design will be the more economical.

Codes of practice always permit elastic design but may not always permit ultimate load design without certain limitations. Because of the

Table 6.1 ULTIMATE LOAD AND ELASTIC PROPERTIES OF UNIVERSAL BEAM AND CONCRETE SLAB COMBINATIONS[10]

Steel beam	Concrete slab Thickness (in)	Breadth (in)	3000 Resistance moment (ton in)	3000 Concrete compression (tons)	3000 Neutral axis depth (in)	4000 Resistance moment (ton in)	4000 Concrete compression (tons)	4000 Neutral axis depth (in)	6000 Resistance moment (ton in)	6000 Concrete compression (tons)	6000 Neutral axis depth (in)	m=15 Inertia moment (in)	m=15 Section modulus Concrete (in)	m=15 Section modulus Steel (in)	m=30 Inertia moment (in)
$14 \times 6\frac{3}{4} \times 34$	5	84	2231	230·0	4·60	2363	230·0	3·45	2495	230·0	2·30	1063	3188	75·9	895
$I = 339.2$		72	2145	214·3	5·05	2297	230·0	4·03	2451	230·0	2·68	1026	2908	74·9	895
$Z = 48.5$		60	2052	178·6	5·17	2205	230·0	4·83	2390	230·0	3·22	983	2601	73·7	811
$S = 54.5$		48	1955	142·9	5·28	2084	190·5	5·13	2297	230·0	4·03	928	2262	72·2	757
		36	1854	107·1	5·40	1955	142·9	5·28	2145	214·3	5·05	856	1884	70·3	690
		24	1736	71·4	6·81	1819	95·2	5·43	1955	142·9	5·28	757	1460	67·5	605
	4	72	1947	171·4	4·19	2067	228·6	4·00	2221	230·0	2·68	897	2649	69·5	749
		60	1883	142·9	4·28	1988	190·5	4·13	2160	230·0	3·22	859	2359	68·5	710
		48	1817	114·3	4·37	1905	152·4	4·25	2067	228·6	4·00	811	2045	67·3	664
		36	1747	85·7	4·73	1817	114·3	4·37	1947	171·4	4·19	749	1704	65·6	608
		24	1639	57·1	6·89	1718	76·2	5·45	1817	114·3	4·37	664	1330	63·1	540
	3	48	1705	85·7	3·73	1760	114·3	3·37	1862	171·4	3·19	700	1800	62·7	577
		42	1676	75·0	4·54	1733	100·0	3·42	1825	150·0	3·26	675	1655	62·1	556
		36	1639	64·3	5·35	1705	85·7	3·73	1786	128·6	3·33	647	1507	61·3	533
		30	1594	53·6	6·16	1665	71·4	4·81	1746	107·1	3·40	615	1353	60·3	508
		24	1540	42·9	6·97	1610	57·1	5·89	1705	85·7	3·73	577	1195	59·2	481

Column group headings: "Ultimate load values for steel to BS 968 for concrete cube strength u_w (lbf/in²)" spans the 3000, 4000 and 6000 groups. "Elastic properties for modular ratio m" spans the m=15 and m=30 groups.

Note. Values in Imperial units — for illustration only.

high shape factor of the composite section, working load stresses may be very close to ultimate stresses. CP 117 for example, gives limits for working load stresses in a composite beam designed by ultimate load methods of 0·9 of the yield stress in steel and 0·33 of the cube strength in concrete.

Universal beam design tables

Tabulated values of relevant section properties of concrete slabs acting compositely with steel beams are available from a number of sources.[9, 10, 11] A typical set of data is shown in *Table* 6.1 for one beam and slab combination. The data have been presented in such a way that design to CP 117 part 1 on a load factor basis is facilitated. This code requires a check to be made of working load stresses, which must not exceed specified values. Where there is a possibility that the values may be exceeded, the tabular entries have been set in italic face type.

In the event of composite beam tables not being available, an initial estimate of the steel beam section modulus required may be found by dividing the total bending moment by the allowable steel stress increased by the factor shown in *Table* 6.2

Table 6.2 FACTOR FOR INCREASE IN STEEL STRESS WHEN ESTIMATING COMPOSITE STEEL BEAM SIZE[8]

m	Factor
15	1.2
12	1.24
10	1.28
8	1.33
6	1.37

To calculate the ultimate moment of resistance of a composite beam it is necessary to specify the following parameters:

Concrete cube strength	f_{cu}
Slab thickness	d
Slab effective width	b
Steel yield stress	f_y
Steel beam geometry	

(It is assumed that the slab rests directly on the steel beam without a haunch).

For a given steel beam (geometry and f_y fixed) and a slab of given thickness the ultimate moment of resistance is a function of the product of cube strength and effective width. Design charts for the range of universal beams giving ultimate moments of resistance related to the product of cube strength and effective width for a given slab thick-

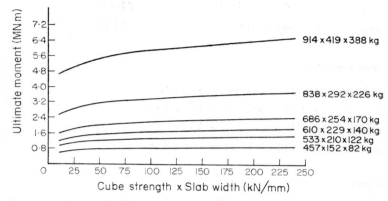

Figure 6.4 Ultimate moment of resistance of composite beam. Universal beams composite with concrete slabs–thickness 125 mm

ness have been published[9]. A typical chart of this type is shown in Figure 6.4. Rapid selection of beam size is made possible in this way, subject to the usual working load stress and deflection checks.

6.7 Design examples

The examples are based on a common beam geometry and loading. In this way the variations in steel weight consequent upon different design approaches are clearly illustrated.

General data

Beam span	9·150 m
Beam spacing	3·810 m
Slab	127 mm thick
Floor finishes etc.	1 kN/m²
Assumed beam weight	0·875 kN/m
Imposed (live) load	5 kN/m²

Loads

Steel beam	$(= 0.875 \times 9.15)$	$= 8.0$ kN
Slab	$(= 3.81 \times 0.127 \times 9.15 \times 23.5)$	$= 105$ kN
Finishes	$(= 3.81 \times 9.15 \times 1)$	$= 34.9$ kN
Imposed	$(= 3.81 \times 9.15 \times 5)$	$= 174$ kN

Bending moments

Steel beam	$(= 8.0 \times 9.15/8)$	$= 9.15$ kN m
Slab	$(= 105 \times 9.15/8)$	$= 120$ kN m
Finishes	$(= 34.9 \times 9.15/8)$	$= 40$ kN m
Imposed	$(= 174 \times 9.15/8)$	$= 199$ kN m

Summary

Construction	Bending Moments (kN m)	
	Steel Beam	Composite Section
Unpropped	129	239
Propped	—	368

Effective width

Least of:

One-third beam span   3.05 m
Distance between beams   3.81 m
Twelve times slab thickness plus breadth of rib, say, 1.524 m
(breadth of rib neglected)

Adopt 1.524 m as effective width.

Materials

Concrete

Type	A	B	C
28 day cube strength f_{cu}	21	28	42 N/mm^2
Allowable bending stress	7	9.3	14 N/mm^2

Steel

	Grade 50	Grade 43
Yield stress	355	250 N/mm²
Allowable bending stress	230	165 N/mm²

Deflection

Limiting the live load deflection to 'span/360' requires a minimum second moment of area

$$I = \frac{5 \times 360}{384 \times E} WL^2 = 4 \cdot 68 \frac{WL^2}{E}$$

$$\left(= \frac{4 \cdot 68 \times 209 \times 9 \cdot 15^2}{2 \cdot 1 \times 10^5} \times 10^3 \times 10^6 \right) = 3 \cdot 9 \times 10^8 \text{ mm}^4$$

EXAMPLE 6.1   STEEL BEAM

For comparison a size of steel beam to carry all loads without composite action is calculated.

Grade 43

Section modulus $(= 368 \times 10^6 / 165) = 2 \cdot 23 \times 10^6 \text{ mm}^3$

Adopt 533 mm × 210 mm × 101 kg/m universal beam

$S_s = 2 \cdot 29 \times 10^6 \text{ mm}^3$

$I = 6 \cdot 15 \times 10^8 \text{ mm}^4$

Grade 50

Section modulus $= 368 \times 10^6 / 230 = 1 \cdot 6 \times 10^6 \text{ mm}^3$

Adopt 533 mm × 210 mm × 82 kg/m universal beam

$S_s = 1 \cdot 79 \times 10^6 \text{ mm}^3$

$I = 4 \cdot 73 \times 10^8 \text{ mm}^4$ (Deflection criterion rules)

6.7.1 UNPROPPED COMPOSITE ACTION

The dead load of beam and slab will be carried on the steel beam acting alone. The final steel stresses will be a combination of non-composite and composite stresses, the concrete stresses being caused solely by composite loading. A direct design solution using tabulated values is not possible but a good starting point can be obtained by evaluating the minimum acceptable concrete section modulus (composite bending moment divided by the permissible concrete stress). From tabular values a section may then be selected which is acceptable on a basis of concrete stress and minimum second moment of area to limit deflection. For elastic designs the remaining criterion is restriction of steel working stresses to allowable values; in addition, for ultimate load designs, the ultimate moment of resistance of the composite section must be adequate to carry the ultimate bending moment.

Initial design criteria for unpropped construction

Minimum concrete section modulus $(= 239 \times 10^6/7) = 3 \cdot 41 \times 10^7$ mm³
Minimum composite second moment of area $= 3 \cdot 9 \times 10^8$ mm⁴
Minimum ultimate moment of resistance $(= 1 \cdot 75 \times 368) = 644$ kN m
   It should be noted that, where a fixed value of modular ratio is used for calculating the section properties (as is the case in these examples), the composite elastic moduli are independent of the grade of concrete used in the slab, and depend only on the *size* of the slab.

EXAMPLE 6.2   ELASTIC DESIGN

Steel grade	Beam	Weight (kg/m)	I_t $\times 10^8$	S_s $\times 10^6$	S_1 $\times 10^7$	S_4 $\times 10^6$
43	533×210×92	92	12·4	2·07	8·4	2·82
50	457×191×67	67	7·35	1·29	6·24	1·85

The values I_t and S_1 exceed those required for deflection and concrete stress limitation. It remains therefore to check the steel stress in the bottom flange

$$\text{Grade 43}\quad (129 \times 10^6/2 \cdot 07 \times 10^6) = 62 \cdot 4 \text{ N/mm}^2$$
$$(239 \times 10^6/2 \cdot 82 \times 10^6) = 84 \cdot 8 \text{ N/mm}^2$$
$$\text{Total}\quad 147 \cdot 2 \text{ N/mm}^2$$

Grade 50 $(129 \times 10^6/1 \cdot 29 \times 10^6) = 100$ N/mm^2

$(239 \times 10^6/1 \cdot 85 \times 10^6) = 129$ N/mm^2

Total   229 N/mm^2

Thus in both cases the steel stresses are below the allowable values.

Using the initial design factor for $m = 15$ (Table 6.2) the required steel section moduli may be estimated as

Grade 43 $S_s (= 368 \times 10^6/165 \times 1 \cdot 2) = 1 \cdot 86 \times 10^6$ mm^3

Grade 50 $S_s (= 368 \times 10^6/230 \times 1 \cdot 2) = 1 \cdot 33 \times 10^6$ mm^3

and it will be found that these values are close to those actually selected.

Where the allowable steel stress in only marginally exceeded a single central prop may be used to reduce the initial stress.

EXAMPLE 6.3   ULTIMATE LOAD DESIGN

Although the concrete strength has no effect on the elastic section properties (m constant) it does affect the ultimate moment of resistance of the composite section. The values below have been selected from composite beam tables.

Steel grade	Concrete f_{cu}	Beam mm $\times$ mm $\times$ kg/m	M_u (kNm) $\times 10^2$	I_t $\times 10^8$	S_s $\times 10^6$	S_1 $\times 10^7$	S_4 $\times 10^6$
43	21	457 $\times$ 191 $\times$ 74	8·7	8·18	1·46	6·56	2·06
	42	457 $\times$ 191 $\times$ 67	9·05	7·35	1·29	6·24	1·85
50	21	406 $\times$ 178 $\times$ 60	6·54	5·72	1·06	5·30	1·54
	42	356 $\times$ 171 $\times$ 57	6·62	4·50	0·894	4·49	1·36

Inspection shows that M_u (moment of resistance), I_t and S_1 exceed the minimum values required. Check steel stresses–lighter sections in each grade.

Grade 43

$(129 \times 10^6/1 \cdot 29 \times 10^6)$ $= 100$ N/mm^2

$(739 \times 10^6/1 \cdot 85 \times 10^6)$ $= 129$ N/mm^2

Total   229 N/mm^2

Allowable stress $(= 250 \times 0 \cdot 9)$ $= 225 \cdot 0$ N/mm^2

There is a small overstress here. It will be found that sufficient additional capacity is available if the concrete width is increased by 75 mm. This may be done by adding the steel beam flange width to the effective width originally considered.

Grade 50

$(129 \times 10^6 / 0.894 \times 10^6)$		$= 144 \text{ N/mm}^2$
$(239 \times 10^6 / 1.36 \times 10^6)$		$= 175 \text{ N/mm}^2$
	Total	319 N/mm^2
Allowable stress	$(= 355 \times 0.9)$	$= 320 \text{ N/mm}^2$

6.7.2 PROPPED COMPOSITE ACTION

The total load is carried by the composite section so that the complication of evaluating the steel beam stress in two stages now disappears. It is possible, in tables based on CP 117 design rules, to indicate whenever the ultimate moment of resistance of the composite beam has a magnitude such that the allowable working stress in concrete f_1 is exceeded; writing the concrete working stress $\sigma_1 = M/S_1$

we also have
$$M_u = 1.75M$$

therefore
$$M_u = 1.75\sigma_1 S_1$$

but σ_1 must not exceed some value f_1:

$$M_u \leq 1.75 f_1 S_1$$

if the working stress is not to exceed f_1.
For elastic design the required section moduli can be calculated directly from the allowable stresses.

EXAMPLE 6.4   ELASTIC DESIGN

Using the allowable stress values the required section moduli for a bending moment of 368 kN m are:

	$(mm^3 \times 10^7)$
S_1 f_{cu} 21 N/mm²	5·26
42 N/mm²	2·63
S_4 Grade 43 steel	0·221
Grade 50 steel	0·160

from which criteria the following steel beams are selected.

Grade 43   457×191×82 Universal beam

Grade 50   406×178×67 Universal beam

EXAMPLE 6.5   ULTIMATE LOAD DESIGN

By direct selection from tables

Steel	Concrete f_{cu}	Beam (mm × mm × kg/m)
Grade 43	21 N/mm²	457×191×74
	42 N/mm²	457×152×67
Grade 50	21 N/mm²	406×178×60
	42 N/mm²	356×171×57

Summary

Table 6.3 contains an analysis of the steel beam weights required for the various conditions investigated above, from which it is clear that ultimate load design of propped construction in high yield steel and

Table 6.3 COMPARISON OF BASIC BEAM WEIGHTS

Steel	Design method	Concrete f_{cu} 21 N/mm²		Concrete f_{cu} 42 N/mm²	
		Unpropped	Propped	Unpropped	Propped
Grade 43	Steel only	100	100	100	100
	Elastic	91	81	91	81
	Ultimate	74	74	66	66
Grade 50	Steel only	81	81	81	81
	Elastic	61	61	61	61
	Ultimate	59	59	57	57

high strength concrete produce the greatest saving in steel. It should be noted that the lightest composite section has a second moment of area nearly three times as large as its steel beam element; deflection is therefore not likely to be critical on average spans. However, its value is close to the limit for deflection and so further reduction in steel beam size may be curtailed, even if the working stress limits were raised.

Although ultimate load design produces the greater saving in steel weight when compared with elastic design, propping does not lead to still greater economy because the ultimate moment is not affected by any placing or removal of supports.

EXAMPLE 6.6 COMPOSITE BEAM WITH RIBBED STEEL SHEETING

For comparison, the effect of steel sheet having a rib height of 37 mm with the ribs running at right angles to the beam span is calculated. Considering propped construction and the following beam (Figure 6.5)

	A_s	D	b_4	t_w	t_4
457×191×74kg/m	9490	457	191	9·1	14·5

$f_{cu} = 21$ N/mm² $\quad f_y = 250$ N/mm²

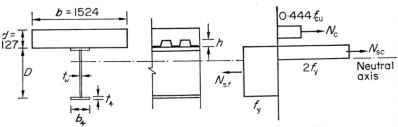

Figure 6.5 Composite beam with ribbed sheet–Example 6.6

the neutral axis depth will be below the slab if

$$A_s f_y > bd\, 0·444$$

$$A_s f_y (= 9·49 \times 2·5 \times 10^5) = 2·38 \times 10^6 \text{ N}$$

$$0·444bd\, f_{cu}(= 0·444 \times 1·524 \times 1·27 \times 2·1 \times 10^6) = 1·81 \times 10^6 \text{ N}$$

Thus the neutral axis lies below the top of the sheeting ribs and the depth of slab for composite action is limited to $(127 - 37) = 90$ mm.

From equilibrium (Figure 6.5),

$$N_c = 0·444 \times 1·524 \times 9·0 \times 2·1 \times 10^5 = 1·28 \times 10^6 \text{ N}$$

$$N_{st} = 2·38 \times 10^6 \text{ N}$$

$$N_{sc} = 2A'_s \times 250 = N_{st} - N_c[= (2·38 - 1·28)10^6] = 1·10 \times 10^6 \text{ N}$$

$$A'_s = (1·1/5·0) \times 10^4 = 2200 \text{ mm}^2$$

Neutral axis lies $2200/191 = 11·5$ mm below top of top flange.

Taking moments of forces about the steel neutral axis

M_u (1280×0·311) = 398 kN m

  (1110×0·223) = 247 kN m

  Total $\overline{645}$ kN m

The required ultimate moment of resistance is 644 kN m so that the beam still just meets the ultimate moment requirement. However, its elastic properties have been changed by the loss of concrete. The effect on S_4 is very small and can be neglected but the change in S_1 is more significant. It may be calculated approximately from

$$S_{1\ rib} = (1 - h/2d)S_1$$

In this particular case

$$(1 - h/2d) = 1 - 37/2 \times 127 = 0·854$$
$$S_{1\ rib}(= 0·854 \times 6·56 \times 10^7) = 5·6 \times 10^7 \text{ mm}^3$$

which still exceeds the minimum value required.

The change in the second moment of area of the composite beam is given approximately by

$$I_{t\ rib} = (1 - h/5d)I_t$$

and this value should be used in deflection calculations.

6.8 Encased beams

Where, for some reason, concrete encasement of steel beams is necessary, composite action without the use of shear connectors may be assumed provided the following criteria are satisfied:

Minimum concrete strength 21 N/mm² at 28 days

Minimum width of casing $(B + 100)$ mm where B is the breadth of the steel beam

Minimum cover to surfaces and edges of flanges 50 mm

Stirrups to be at least 5 mm in diameter at not more than 150 mm centres and symmetrically placed (Figure 6.6)

Design is permitted only by elastic methods by CP 117, and then only if heavy concentrated loads are *not* carried.

6.9 Shear connection

Under working load conditions the horizontal shear force distribution may be found by a conventional elastic analysis. However, in buildings, in which moving and varying loads are not generally present, a simpler method is to calculate the maximum horizontal force at ultimate load at the beam–slab interface (which occurs at the point of maxi-

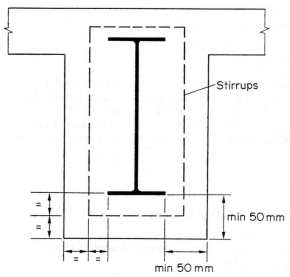

Figure 6.6 Encased beam

mum bending moment) and to provide sufficient shear connectors between this point and the point of zero bending moment to transfer the force.

For uniform loading the connectors may be spaced uniformly; where heavy concentrated loads occur the distribution may be uniform between points of discontinuity of the shear force diagram but the number of connectors between these points must be proportional to the area of the diagram between the points (see Figure 6.7).

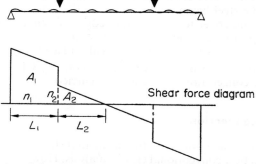

Figure 6.7 Spacing of shear connectors for concentrated loads. The total number N of connectors between points of zero and maximum bending moment is given by

$$N = n_1 + n_2$$
$$n_1 = N[A_1/(A_1 + A_2)] \qquad n_2 = N[A_2/(A_1 + A_2)]$$

The interface force is equal to the axial force in the slab if the neutral axis lies within the beam, or the axial force in the beam if the neutral axis lies in the slab.

EXAMPLE 6.7 DESIGN OF SHEAR CONNECTORS

For the composite sections designed in Examples 6.1 to 6.5 the plastic neutral axis falls within the steel beam. The ultimate concrete force is therefore the force to be transmitted by the shear connectors, which for concrete of strength 21 N/mm² will be

$$0.444 \times 21 \times 1.524 \times 127 \times 10^3 \times 10^{-3} = 1810 \text{ kN}$$

From *Table* 4.1 the design value (80 % of ultimate capacity) of 16 mm diameter, 76 mm long, headed studs in concrete of strength 21 N/mm² is 57 kN. Placing the studs in pairs the number of pairs over the half length of the beam is $1810/2 \times 57 = 16$.

Uniform spacing will result from a centre-to-centre distance of 286 mm between pairs starting 143 mm in from each end of the beam. The code limitation on spacing is the smaller of

Four times slab thickness $(4 \times 127 =)$ 508 mm

or a fixed value 610 mm

which in this case does not apply.

For the beam with ribbed sheeting of Example 6.6 the ratio of rib width to height, w/h, is 1·5. The shear connector allowable load is

$$0.5 \times 1.5 \times 57 = 42.8 \text{ kN}$$

The force to be transferred is 1280 kN, giving the number of pairs over the half length of the beam as

$$1280/2 \times 42.8 = 15$$

Transverse reinforcement

To avoid failure in longitudinal shear around the connectors the shear stress on a plane of length L_s is limited to specified values. L_s is defined as the peripheral distance round the shear connector group or twice the slab thickness, whichever is smaller. The shear force per unit length of beam v, which is the design value of the shear connectors divided by their spacing, must not exceed the lesser of the values of the shear resistance per unit length of beam given by (units Newtons and millimetres)

(a) $0 \cdot 234 \, L_s \, (f_{cu})^{1/2} + A_b f_y n$

(b) $0 \cdot 625 \, L_s \, (f_{cu})^{1/2}$

where
A_b is the area, per unit length of beam, of transverse reinforcing bars in the bottom of the slab, n is the number of times each transverse reinforcing bar in the bottom of the slab is intersected by the shear plane (for T beams $n = 2$, for L beams $n = 1$ generally).

The value of f_y is the yield stress of the reinforcing bars but must not exceed 420 N/mm².

Additionally a minimum area of transverse bottom reinforcement must be provided (steel already present for transverse bending may be counted in this area) at the rate of $(v/4f_y)$ mm²/mm.

Applying these requirements to the beam under consideration

$$v(= 2 \times 57 \times 10^3/286) = 3 \cdot 99 \times 10^2 \text{ N/mm}$$
$$(u)^{1/2}[= (21)^{1/2}] = 4 \cdot 58 \text{ N/mm}^2$$
$$L_s(= 2 \times 127) = 254 \text{ mm (less than periphery of connectors)}$$
$$A_t(= 3 \cdot 99 \times 10^2/4 \times 303) = 0 \cdot 33 \text{ mm}^2/\text{mm}$$

Adopt 8 mm diameter bars at 150 mm centres.

$$0 \cdot 234 \;\; L_s(u)^{1/2} + A_t f_y n = 0 \cdot 234 \times 254 \times 4 \cdot 58 + 0 \cdot 33 \times 303 \times 2$$
$$= 473 \text{ N/mm}$$
$$0 \cdot 625 \, L_s(u)^{1/2} \, (= 0 \cdot 625 \times 254 \times 458) = 726 \text{ N/mm}$$

Both these criteria exceed the actual value, 399 N/mm, of v.

REFERENCES CHAPTER 6

1. CREASY, L. R., Composite construction,' *Struct. Engr*, **42,** No. 12 (Dec. 1964)
2. DORMAN, A. P., FLINT, A. R., and CLARK, P. J., Structural steelwork for multi-storey building — design for maximum worth', *Proc. Conf. on Steel in Architecture*, British Constructional Steelwork Association, London (1969)
3. *Modern fire protection for structural steelwork*, British Constructional Steelwork Association, Pamphlet FP3 (Nov. 1967)
4. ROBINSON, H., 'Tests on composite beams with cellular deck', *Proc. Am. Soc. civ. Engrs Struct. Div.*, **93** (Aug. 1967).
5. ROBINSON, H., *Proposed method of establishing design criteria for composite steel and concrete beams incorporating cellular steel decking*, McMaster University, Hamilton, Ontario (1964)

6. *Composite beam manual for the design of steel beams with concrete slab and cellular steel floor*, Canadian Sheet Steel Building Institute, Port Credit, Ontario (1968)

7. FISHER, J. W., 'Design of composite beams with formed metal deck', *Am. Inst. Steel Constr. Engng J.*, **7**, No. 3 (July 1970). Discussion in **8,** No. 1 New York (Jan. 1971)

8. LOTHERS, J. E., *Advanced design in structural steel*, Prentice-Hall, New York (1960)

9. DAVIES, C., Design Charts for composite beams,' *Civ. Engng publ. Wks Rev.* London (May 1967)

10. *Composite construction for steel framed buildings*, British Constructional Steelwork Association, London, Publication No. 25 (1965)

11. *Manual of steel construction,* American Institute of Steel Construction New York (1966)

CHAPTER 7

Composite Bridges

7.1 History

Steel beams supporting concrete slabs have been used to form the basic superstructure of large numbers of deck bridges for many years but the precise date on which an enterprising engineer first made a decision to ensure positive shear connection between beam and slab in the design of these bridges is not known. During the early part of the decade 1930–40 composite bridges were being built in the relatively remote island of Tasmania, which leads to the supposition that others were being built elsewhere (Chapter 1, reference 8). However, it was not until after 1945 that, for various reasons (economics being one), the number of composite bridges being built annually showed a significant increase. Reference has already been made to the stimulus given to economic forms of construction by the steel shortages in Germany immediately after the end of the 1939–45 War. New codes of practice in other countries, the publication of papers describing the results of experimental work and eventually the publication of text books have all helped to make engineers familiar with composite construction; today there would have to be some special reason for specifying a non-composite concrete decked steel beam bridge in preference to its composite counterpart.

The state of the art of composite bridge design in Australia in 1934 is apparent from the paper by Knight (Chapter 1, reference 8), which refers to the replacement of wooden bridges throughout Tasmania by permanent structures. It was realised that steel beams supporting concrete deck slabs were an economical solution particularly if composite action could be achieved. The arguments for composite action were principally those advanced in Chapter 1 relating to the more effective use of the concrete in the deck slab, though the ability to use smaller steel beams for a given span must have been of importance in an island into which all steel sections had to be imported. It is also very interesting to see that the method of prestressing the steel beams by jacking up at intermediate points was described and that considerable space was devoted to the problem of finding the bending moments in continuous concrete slabs supported by elastic beams—the load distribution problem. An experimental investigation of load distribution was carried out, a design example was given and certain other fundamental points (modular ratio, effective width, shear connection) were discussed. In fact the contents of this paper could be used today as the basis for a composite bridge design method.

7.2 Economics

In the field of bridge design the engineer has the greatest scope for exercising his skill. The wide variation in spans, widths, locations and construction problems lead to an equally great variety of solutions, of which the composite deck is only one. Nevertheless the majority of bridges lie in the short (up to about 30 m) span range, for which a beam and slab solution is economic. The designer's choice is then generally between providing concrete (reinforced or prestressed) beams or steel beams to support the concrete deck. The notorious difficulty of assessing the relative cost of different types of construction on an impartial basis is well known. All that need be said here is that composite bridge decks are generally competitive and indeed have certain specific advantages, notably a lower dead weight, when compared with their concrete counterparts in the short span range. An illustration of relative weights is given in *Table* 7.1 for composite steel and concrete prestressed beams with in situ infill and prestressed beam and slab bridges. Relative costs are also given but these must be treated with caution since they apply only to a specific country and date.[1]

7.3 Composite bridges in Europe

Amongst general trends which have been noticeable in European bridge design in recent years there are three of particular interest:

(a) The increasing use of box section girders
(b) The virtual disappearance of riveted construction and its replacement by welding and high strength friction grip bolting
(c) The adoption of higher strength steels[2]

The tubular (round or rectangular) member has a long history in civil engineering, a classical example in bridges being Stephenson's Britannia Bridge, opened in 1850.[3] However, the fabrication of box members involving the shaping and jointing of plates, was not easy and when rolled steel beams became available the box section was to a large extent superseded by them. Recent advances in steel fabrication technology have once more focused designers' attention on the merits of the box girder. Steel plate can readily be cut to complicated shapes by gas cutting and the resultant pieces joined by electric arc welding without difficulty. Both processes can, in the right circumstances, be performed automatically. The resulting beams are structurally more efficient and, because of their smooth flat surfaces, less liable to corrosion, easier to maintain, and of better appearance than their conventional 'I' girder counterparts.

Table 7.1 RELATIVE COST OF SHORT SPAN BRIDGES

| | \multicolumn{6}{c}{*Span* (m)} |
| Bridge type | \multicolumn{2}{c}{*12*} | \multicolumn{2}{c}{*18*} | \multicolumn{2}{c}{*24*} |
	Weight	*Cost*	*Weight*	*Cost*	*Weight*	*Cost*
Composite steel and concrete	40·2	19·8	43·3	23·7	57·6	31·7
Prestressed beam in situ infill	134·0	21·0	146·0	29·2	—	—
Prestressed beam and slab	—	—	—	—	125·0	38·0

Weights in tonne/m² of deck
Costs in £/m² of deck

Steelmaking technology has made available steels of enhanced strength and, at the same time, has found ways of improving the metal's resistance to the danger of the sudden catastrophic failure known as 'brittle fracture', which was particularly prevalent in welded steel structures. A combination of the correct type of steel (having adequate 'notch ductility'), the correct welding technique and careful detailing will ensure that brittle fracture is not a danger. Finally, high strength friction grip bolts have led to more compact joints more easily made in situ.

The bridges described in the following pragraphs have been chosen in order to demonstrate a variety of design and construction techniques adopted for composite bridges in Europe.[4]

7.3.1 THE RAITH BRIDGE, SCOTLAND

The problems of continuous composite construction have generally been avoided in Great Britain and indeed it is believed that the Raith Bridge is the first example in that country in which a deliberate attempt

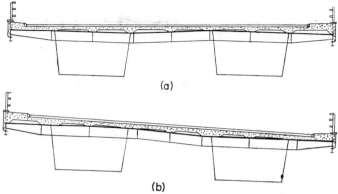

(a)

(b)

Figure 7.1 The Raith Bridge. (a) Typical part cross-sections at midspan; (b) Typical part cross-sections at piers

has been made to provide a compressive prestress in the concrete slab by an erection technique. The girders are continuous trapezoidal box sections from which most of the top plate has been omitted: the closure to the box is provided by the slab, which acts compositely through narrow top flange plates welded to the web plates of the boxes. The arrangement is an efficient one because in conventional symmetrical girder systems the top flange stresses are generally well below the optimum value and some reduction in top flange area is a possibility.

The jacking method was used to prestress the deck. The continuous box beams were first rolled out into position. The level of the intermediate supports was maintained some 0·5 m above the end supports while the deck was cast and allowed to harden. Then the intermediate supports were jacked down to level.

The bridge carries twin three-lane carriageways on continuous spans of $42·6 + 51·8 + 42·6$ m. Relatively thin plates are used for the trapezoidal boxes, the webs being only 10 mm thick, and so both longitudinal and transverse stiffening is required for webs and bottom flanges.

7.3.2 WEIBERSWOOG BRIDGE, NAHETAL, GERMANY

In Germany many continuous bridges have been built in which, by prestressing, the concrete slab is always in compression along the whole length of the bridge under all conditions of loading. This five-span continuous bridge $(57·8 + 72 + 72 + 72 + 57·8$ m) is partly straight, partly curved in plan. The steelwork is an open trough section of stiffened plate. To avoid prestressing the complete girder the top flanges of the trough are composed of two plates, the upper (carrying the shear connectors) being free to slide over the lower. After the deck was cast

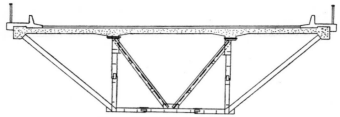

Figure 7.2 Weiberswoog Bridge

it was post-tensioned with longitudinal and transverse cables. When an appropriate time had elapsed, during which much of the total creep and shrinkage took place, the sliding plate was welded to the flange plate below.

7.3.3 PONT DE LA MADELEINE, FREIBURG, SWITZERLAND

Precast concrete deck units have not been widely used although there are examples in Germany and Switzerland of this type of construction. The advantages of speed and absence of shuttering, which may be obtained by the use of precast units, are particularly of value in the construction of footbridges and elevated highways over busy streets. Composite action is most simply obtained by forming holes in the concrete units into which the shear connectors project. At the appropriate time in the construction sequence the holes are grouted up. Alternatively the slabs may be cast onto suitable steel sections, the latter then being bolted or welded to the main steelwork.

Pont de la Madeleine is a three-span continuous structure (85·5 + + 106·5 + 85·5 m) comprising two high-tensile steel plate girders

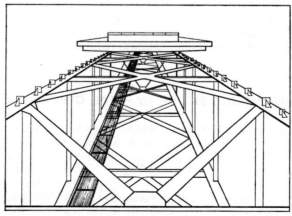

Figure 7.3 Pont de la Madeleine, Freiburg

4·212 m deep and 5·900 m apart. The girders were launched as a continuous beam, the internal bracing was fixed and then the ends over the abutments were lowered to final level while the temporary supports over the piers were maintained 1·6 m above final level. The precast deck units, each 11·050 m wide, 2·000 m long and 240 mm thick were then placed in position. 'T' shear connectors at 1·000 m centres were linked to each slab by a reinforcement bar protruding through each pocket in the concrete.

The deck for a length of 24 m each side of the two piers was prestressed with cables, the sequence of operations being

(a) insert cables in ducts
(b) fill transverse joints between slabs to be prestressed with concrete
(c) tension cables
(d) fill remaining transverse joints and shear connector pockets with concrete

After all the infilling concrete had hardened the girders were lowered over the piers to final level.

7.3.4 TAY ROAD BRIDGE

Mention has been made in Chapter 3 of the way in which part of the Tay Road Bridge deck was cast while the steel box beam was continuously supported by the ground: a 55 m span box beam at this stage weighed about 1800 kN and, as a composite girder, could carry the full dead load of the superstructure. The complete bridge consists of 42 spans, of which all but the four longest (the navigation spans) are simply supported. Such a large number of relatively short spans is generally an economical solution where foundation conditions for the piers are good.

The four-span continuous section $(70+76+76+70$ m$)$ is of box beams of varying depth. The jacking method of prestressing the deck in the continuous section was rejected because the jacking lift of about 3·0 m was considered excessive. Instead a combination of phased casting and extra reinforcement was used, the deck over the intermediate supports being placed last. In this way the composite negative dead load moment was reduced as far as possible. At the points of contraflexure, joints were made in the deck slab (Figure 7.4).

7.4 Bridge Design

Many complex decisions are required of a bridge designer before he can determine the structural system to be employed in a particular bridge scheme. A number of alternative schemes will inevitably have to be considered, at least in outline, and comparative costs evaluated.

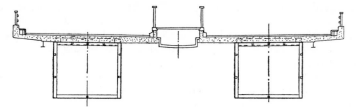

Figure 7.4 The Tay Bridge

To a great extent the decisions will be based on the engineer's accumulated experience aided by available published material. With notable exceptions[5] the available published material is scattered among many books and journals and, for a bridge of any size, considerable research will be required to locate relevant reference sources.

In this section it is assumed that, as a result of consideration of all factors, a decision has been made to use some form of composite construction for the bridge superstructure and that certain preliminary points concerning the dimensions (girder spacing, deck thickness and the like) have been established. This first *synthesis* then has to be followed by a detailed *analysis* of the system in order to establish whether it is structurally sound. In many ways the initial synthesis is the most difficult part of the design process, demanding a considerable degree of background knowledge in the engineer. However, for the simpler bridges, fairly accurate guesses may be made with the help of the design aids mentioned below. Analysis of the system is considerably speeded by the use of the electronic computer. Semi-automatic design of standard bridges, for example motorway overbridges, is now a possibility and, because of the speed of the computer's operation, analysis of a number of alternative designs is greatly facilitated.

While initial estimates of dead loading depend on the designer's ability to synthesise a suitable cross section, the live loading is generally 'accurately' known from the beginning. In fact, estimates of live loads on bridges are almost as difficult to make as those for buildings; with the added complications that the dynamic characteristic and repetitive nature of vehicle loading make to the problem. Most codes of practice simplify the real problem to some extent by the use of equivalent line and uniformly distributed loads designed to simulate the effects of some realistic train of heavy axles. In addition the problem of the abnormally heavy vehicle is covered by specifying the dimensions and weight of this vehicle. The dynamic effect may be catered for by an impact factor to be applied to the static load. The time-dependent nature of loading caused by an endless succession of vehicles of varying size crossing a bridge has been a source of difficulty especially as it is necessary to extrapolate to some date, possibly fifty or a hundred years in advance, data collected at the present time.

Nevertheless load spectra are available from which estimates of fatigue effects can be made.[6]

To form a basis for the design examples the British loading system is used. It is fully described in the British Standard[7] and the following brief description is intended only to outline the system.

Type HA, the normal loading, consists of an equivalent uniformly distributed load and a line (knife edge) load. Together they approximately simulate the effect of a train of vehicles of a certain size and weight occupying the bridge. Type HB, the abnormal loading, consists of an arrangement of wheels and axles of specified weight and geometry, simulating the load of heavy tractor and trailer combinations. The extent of the uniformly distributed load, and the position of knife edge load or abnormal vehicle may be varied to produce the worst possible effect on the part of the bridge being considered.

7.5 Load distribution

The economics and even the practicability of bridges often depend on maximum utilisation of the structure. The designer is generally trying to minimise the dead weight of structure to carry a given load and the problem becomes intensified as spans increase. For this reason account needs to be taken of the load-distributing properties of the deck system, particularly where concentrated loads of the HB vehicle type are concerned. Considering for example a point load placed directly over one girder, the distribution problem is that of determining the proportion of the point load transferred to the other girders by cross girders, deck slab or other transversely spanning elements. Much attention has been given to the problem and solution techniques of varying complexity and accuracy exist[8]. Suffice it to say that for the initial design of composite decks a useful method is of the type in which cross girders and transverse deck are replaced by an equivalent elastic medium, while the longitudinal beam and slab remain unaltered. Such a method using a *basic function* analysis has been developed by Hendry and Jaeger and is fully described in their book[9]. Where there is access to a computer, standard grillage programs can be used but unless a large amount of work is involved the 'hand' method will be found to be quick and simple. An example of the application of Hendry and Jaeger's method is given in Example 7.1.

Once the actual bending moment distribution corresponding to a given loading system has been determined it is necessary to evaluate the stresses caused by it in the various parts of the composite section, the process being repeated for other types of loading. For example, one Code of Practice[10] requires the following conditions to be taken into account:

Dead load
Live load

Impact effect
Lurching effect ⎫
Nosing effect   ⎬ for railway loading
Centrifugal force ⎭
Longitudinal force
Wind pressure
Temperature effect
Resistance to movement of expansion bearings
Forces on parapets
Erection forces and effects
Shrinkage effects
Forces and effects due to settlement of supports

Elastic design methods for bridges are almost universally adopted, a fact which reflects the uncertainty surrounding a limit state analysis when fatigue effects may be important. Nevertheless, there is a firm intention in Great Britain to establish limit state design as the basis of a comprehensive bridge design directive. Recent work by Flint and Edwards[11] has shown that collapse analysis of composite systems is feasible and that a significant economy may be expected from the adoption of limit state design. For the bridges investigated by Flint and Edwards the load factors against the attainment of the various possible limit states ranged from 0·69 to 16·2 which clearly illustrates the margin available for economy. Because of the inevitable delay in the publication of a limit state code for bridges the examples are based on elastic design with an ultimate load check against plastic collapse.

7.6 Design aids

The inexperienced engineer can avoid the waste of time caused by having to modify a poor initial estimate of member sizes by the use of published data. Tabulated properties and design charts for a large number of slab and rolled beam combinations are available.[12] From the design charts reproduced in Figures 7.5 and 7.6, universal beam sizes may be selected for given loadings and slab dimensions. Thus the short span range is amply catered for, with the understanding that for spans up to approximately 30 m the rolled beam is almost certainly the most economic choice. This span range may be extended by adopting a universal beam with a tension flange plate.

For longer spans the problem of making an initial assessment is more difficult. The plate or box girder is probably the obvious choice but there are many possible combinations of flange and web types which can be used. A number of different solutions will have to be evaluated in order to find the most economic. By applying certain simplifications to the exact equations for the properties of unsymmetrical composite

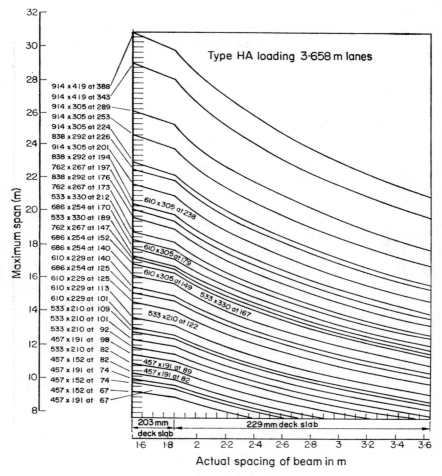

Figure 7.5

plate girders as described in section 2.11 an acceptable initial estimate may be made for simply supported beams.

7.7 Deck slabs

The design of concrete deck slabs spanning transversely across steel beams may be approached in a number of ways:

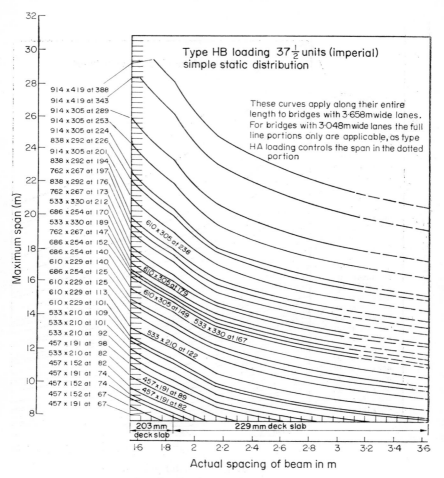

Figure 7.6

(*a*) Consideration of the slab as a series of continuous beam strips on
    rigid supports. This method has the merit of simplicity and,
    provided the deflection of the longitudinal beams is fairly uniform,
    as would be the case for uniformly distributed loading, should
    give acceptable results for the slab bending moments

(*b*) The Hendry and Jaeger method may be used to find transverse
    bending moments in continuous slabs supported by longitudinal
    elastic beams

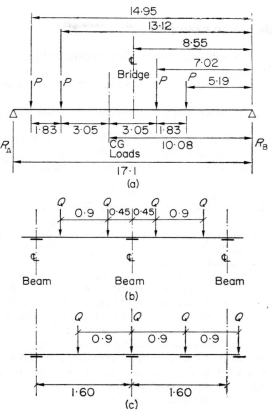

Figure 7.7 Placing of HB *vehicle*

(*c*) Consideration of the slab as a two-way spanning element supported by elastic beams. Clearly such assumptions correspond much more closely to the actual characteristics of the deck system and so should give more accurate results. The necessity to use a computationally more complicated method must depend on the particular deck configuration being analysed. Bending moments in slabs with various edge support conditions may be found by the use of *influence surfaces*,[13] the results being then modified to take account of the elasticity of the slab supports

The final stress distribution in a composite deck slab will be complex since it is a mixture of the effects of longitudinal and transverse bending of the slab alone plus longitudinal bending of the slab and beam combination.

7.8 Bridge design examples

Designs for three different types of composite bridge are outlined in the following pages:

Example 1. A simply supported short span bridge with rolled steel beams

Example 2. A longer simply supported span using welded steel plate girders

Example 3. A continuous three-span box girder bridge

The loading adopted is that of British Standard 153 Part 3A, type HA with a check for the effects of type HB loading in Examples 7.1 and 7.2.

The design of the concrete deck slabs is not considered in detail here. The slab thickness will be dictated by the loading on it; various methods are available for determining the bending moments and shears in concrete slabs supported by flexible beams. Graphical presentation of influence surfaces for plates with varying edge conditions are of value in analysing deck slabs.

It is not suggested that the beam spacings represent an optimum configuration; much will depend on relative costs of steel and concrete.

For full width uniform or knife edge loading on a slab it is possible to determine the loads transferred to the girders using a distribution analysis. However, it will be found in practice that they will not vary much from those calculated assuming the slab to span over the beams as a series of beam strips on simple supports. For HB loading on the other hand, a load distribution analysis will be required to achieve economy. An example of one method, due to Hendry and Jaeger, is included in the load distribution analysis for the short span bridge.

EXAMPLE 7.1 SIMPLY SUPPORTED BRIDGE SPANNING 17 m

General notes

For spans as short as 17 m, a rolled steel section is invariably the most economic beam. A choice between mild or high yield stress steel is available. Consideration should certainly also be given to the use of a corrosion-resistant high yield steel (such as Cor-Ten) which does not require painting: the consequent reduction in maintenance costs may be significant.

The use of tables or charts of the section properties of composite beams will simplify the selection of a trial beam. Indeed the whole design process for such a simple span could be programmed for a computer if the amount of work justified the cost of developing and running the program.

The span range of universal beams may be extended by increasing the bottom flange area with a welded flange plate. The same expedient may be adopted for shorter spans when it is necessary to keep the beam depth to a minimum, as for example when headroom is restricted under the bridge.

Transverse bracing to the compression flange is not required because the concrete deck provides full restraint. Cross girders or diaphragms are generally not required when the beams are at average spacings as the deck itself is an efficient load distributing medium; in cases of doubt a distribution analysis may be made. Although it may not be necessary to provide transverse members, a light form of cross bracing between beams is a useful erection aid.

Economies in deck construction are possible by adopting some form of permanent shuttering. Precast, prestressed planks or plastic coated steel troughed sheeting are two economic alternatives which can replace the more expensive timber shutters. Attention to the details of formwork support can lead to a significant reduction in construction time.

Dimensions and material properties

Simple span	17·1 m
Beam spacing	1·60 m
Slab thickness	200 mm
28 day cube strength	$f_{cu} = 30$ N/mm²
High yield steel	$f_y = 339$ N/mm²

Loadings

Concrete	4·76 kN/m²
Surfacing, etc.	3·1 kN/m²

Live load HA to British Standard 153
Check for 45 Units (Imperial) HB to British Standard 153

Design of girder

The design is based on a typical internal girder. Dead load and HA live load are transferred to each girder in proportion to the deck area it supports. HB load is distributed by a simple method permitted by

BS153 and also by a method due to Hendry and Jaeger. Three loading conditions are investigated:

(a) Steel beam and dead load of slab (non-composite)

(b) Composite (long term) dead load of surfacing

(c) Composite (transient) live load

Loads

Non-composite dead load

Beam (assumed weight)	1·9 kN/m
Slab 4·76×1·6	7·56 kN/m
Total	9·46 kN/m

Maximum bending moment on steel beam alone (M_{gs})

$M_{gs} (= 0\cdot125\times9\cdot46\times17\cdot1^2) = 347$ kN/m

Composite dead load

Surfacing, etc. $(= 3\cdot1\times1\cdot60) = 4\cdot95$ kN/m

Maximum long term bending moment on composite section (M_{gt})

M_{gt} $(= 0\cdot125\times4\cdot95\times17\cdot1^2) = 180$ kN m

Live load

HA The rates of loading appropriate to a span of 17·1 m are:

Uniformly distributed	9·65 kN/m²	(UD)
Knife edge	35·7 kN/m	(KE)

On the basis that each beam supports the loading on the slab area immediately above it the bending moments will be:
Maximum HA bending moment on composite section M_{qt}

$M_{qt}(= 0\cdot125\times1\cdot60\times9\cdot65\times17\cdot1^2) = 579$ kN m

$+0\cdot25\times1\cdot60\times35\cdot7\times17\cdot1 = 244$ kN m

Total   823 kN m

HB The HB vehicle must be placed on the span in the longitudinal and transverse positions that will produce the maximum bending moments in the beams. The longitudinal placing is generally unrestricted but, because of the deck layout (position of kerbs, etc.) there may be restrictions on the transverse placing. In this example it is assumed that the wheels may be put anywhere on the deck.

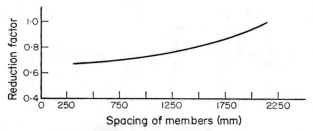

Figure 7.8 Reduction factor for HB *loading*

In the *longitudinal* direction the HB vehicle is placed so that the centre line of the bridge bisects the distance between the centre of gravity of the load and the next wheel (Figure 7.7(a)). This placing generally gives the maximum bending moment. In terms of P (the maximum load (kN) on one beam produced by one axle)

$$10\cdot08 \times 4P = 17\cdot1\ R_A$$
$$R_A = 4 \times 10\cdot08P/17\cdot1 = 2\cdot36\ P$$
$$R_B = (4-2\cdot36)P = 1\cdot64\ P$$

Maximum bending moment occurs under the load nearest to the centre line of bridge.

$$M = 7\cdot02 \times 1\cdot64P - 1\cdot83P = 9\cdot67P\ \text{kN m}$$

The *transverse* position of the load which gives the maximum reaction on any beam is generally one or other of the positions shown in Figure 7.7(*b*) or (*c*).

The reaction from each wheel transferred to the beams may be found in three different ways:

(i) considering each span of deck between beams to be simply supported
(ii) considering the deck as a continuous beam strip on rigid supports
(iii) considering the deck as a continuous beam strip on elastic supports

Method (i) has the merit of simplicity but makes no allowance for the load-distributing properties of the deck. A correction may be applied to take account of distribution of load; a typical set of correction factors from BS 153 is shown in Figure 7.8 (Many other codes have similar provisions.)

Methods (ii) and (iii) represent the physical situation more accurately but at the expense of increased calculation.

12

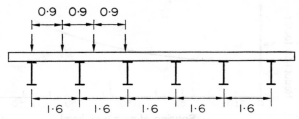

Figure 7.9 Transverse position of HB *vehicle. (Dimensions in* m*)*

(i) Simple statical distribution

Load straddling beam

$$R = \frac{(0\cdot25 + 1\cdot15)}{(1\cdot60)} 2Q = 1\cdot75\,Q$$

Load above beam

$$R = Q + 2Q\frac{(0\cdot70)}{(1\cdot60)} = 1\cdot88\,Q$$

The permitted reduction factor for a beam spacing of 1·60 m is 0·85. The beam will be designed for an HB loading (each axle) of $4Q$ which is equivalent to a longitudinal loading of $P = 1\cdot88 \times 0\cdot85\,Q = 1\cdot6\,Q$ at each loading point.

(ii) Continuous beam on rigid supports

Considering the case in which one wheel lies directly above the beam, the reactions due to the other three wheels may be found from tabulated values for continuous beams:

$$R = Q + 2 \times 0\cdot561\,Q \text{ (approx)} = 2\cdot12\,Q$$

(iii) Continuous beams on flexible supports

If the reactions calculated on the basis of rigid supports as above are applied to the beams, which are then considered flexible, they will deflect different amounts and, because of the load-distributing properties of the deck, the initial reactions will be modified. Analysis requires knowledge of the actual stiffnesses of beams and slab and so will be postponed until a tentative beam design has been made.

As an initial step the simple statical method, modified for transverse distribution is adopted:

$$M = (9\cdot67 \times 1\cdot60)\,Q = 15\cdot5\,Q \text{ kN m}$$

$$M = (15\cdot5 \times 112) \quad = 1740 \text{ kN m} \quad \text{(HB wheel load } Q = 112 \text{ kN)}$$

Section properties

Effective width

The half-width *b* of the slab is 0·80. This is less than one twentieth the beam span (0·875 m) and so the total effective width of the slab is its actual width of 1·60 m.

Modular ratio

For permanent loads $m = \dfrac{83}{(30)^{1/2}} = 15$

For transient loads $m = \dfrac{15}{2} = 7.5$

Try a universal beam $838 \times 292 \times 226$ kg/m nominal size. The relevant section properties for steel beam acting alone and for two types of composite section are shown in *Table* 7.2. Using these values, the stresses caused by each loading case can be evaluated and are shown in *Table* 7.3.

Load distribution analysis

Now that the section properties have been fixed it is possible to perform a load distribution analysis. The method is shown only in outline here; for full details Hendry and Jaeger[9] should be consulted.

The longitudinal beams are considered as fixed supports. It is assumed that there are six beams in the bridge and that the arrangement of wheels shown in Figure 7.9 represents the most severe transverse positioning for each axle. Treating the deck slab as a continuous beam on fixed supports, the reactions on each longitudinal beam caused by the loading may be found using tables of continuous beam reactions or influence lines. The values tabulated below have been

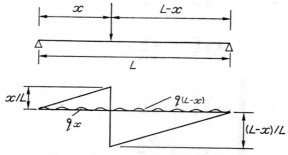

Figure 7.10 Shear force influence line

Table 7.2 SECTION PROPERTIES OF COMPOSITE BEAM OF EXAMPLE 7.1

Composite section

m	A_c (mm² $\times 10^5$)	A_o/m (mm² $\times 10^4$)	A_t (mm² $\times 10^4$)	d_c (mm)	S_t (mm³ $\times 10^8$)	I_t (mm⁴ $\times 10^9$)	y_1 (mm)	$y_2=y_3$ (mm)	y_4 (mm)	S_1 (mm³ $\times 10^7$)	$S_2=S_3$ (mm³ $\times 10^7$)	S_4 (mm³ $\times 10^7$)
7·5	3·2	4·26	7·14	213	9·07	8·17	313	113	738	2·61	7·24	1·11
15·0	3·2	2·13	5·01	302	6·43	6·77	402	202	649	1·69	3·35	1·04

Steel section

A_a (mm² $\times 10^4$)	I_s (mm⁴ $\times 10^9$)	S_s (mm³ $\times 10^6$)
2·88	3·39	7·97

Table 7.3

Stress	M_{gs}	M_{gt}	$M_{q\,HA}$	$M_{q\,HB}$	*Total* HA	*Total* HB
			Stress			
			(N/mm²)			
σ_1	—	− 0·710	− 4·25	− 8·87	−4·96	− 9·58
σ_2	—	− 0·358	− 1·53	− 3·21	−1·89	− 3·57
σ_3	− 43·5	− 5·36	−11·4	− 24·1	−60·3	− 73·0
σ_4	+43·5	+17·6	+74·1	+157·0	+135·0	+219·0

(Tension positive.) For the high yield steel of this example the allowable working stress is 210 N/mm². The infrequent occurrence of the HB vehicle loading justifies an overstress of 4·3 % (the allowance in BS 153 is 25%).

obtained from influence lines. The reactions from a wheel load of Q are first found; these values multiplied by $9·67 \times 112$ give the free bending moments on each beam.

It is now necessary to obtain distribution coefficients for the six-girder bridge. The parameter α which determines the value of the distribution coefficients is given by

$$\alpha = \frac{12}{\pi^4} \frac{(\text{Beam span})^3}{\text{Beam spacing}} \frac{\text{Total transverse stiffness}}{\text{Longitudinal beam stiffness}}$$

The transverse stiffness may be obtained most simply using the second moment of area of the concrete slab calculated ignoring cracking and the contribution of the reinforcement.

Total transverse stiffness (concrete units) $= E_c(17·1 \times 0·2^3)/12$

$$= E_c(1·04 \times 10^{-2})$$

The longitudinal second moment of area is that of a *composite* beam.

Longitudinal stiffness (steel units) $= E_s(8·17 \times 10^{-3})$

With $E_s/E_c = 15$,

$$\frac{\text{Total transverse stiffness}}{\text{Longitudinal beam stiffness}} = \frac{1·04 \times 10^{-2}}{15 \times 8·17 \times 10^{-3}} = \underline{8·48 \times 10^{-1}}$$

$$\alpha = \frac{12}{\pi^4} \frac{(17·1)^3}{1·6} \times 8·48 \times 10^{-1} = 12·6$$

The distribution coefficients corresponding to $\alpha = 12·6$ (the longitudinal beams may be assumed to have zero torsional stiffness) are tabulated in *Table* 7.4. The free bending moments on each beam are then distributed as in *Table* 7.5. The entries are obtained by multiplying the free bending moment by the relevant distribution coefficient.

It is interesting to note that the simple statical method with a reduction factor and the more complicated distribution analysis give almost identical results for this particular bridge.

Table 7.4 DISTRIBUTION COEFFICIENTS FOR SIX-GIRDER BRIDGE

Distribution coefficients	Loaded beam					
	1	2	3	4	5	6
p_1	+0·76	+0·31	+0·04	-0·05	-0·05	-0·01
p_2		+0·37	+0·27	+0·10	+0·01	
p_3			+0·37	+0·27		
p_4				+0·37		
p_5					+0·37	
p_6						+0·76

Note. The distribution coefficients are shown in the form p_{ab} where p is a distribution coefficient giving the proportion of the free bending moment on girder b distributed to girder a.

The reciprocal property and symmetry of the coefficients ($p_{ab} = p_{ba}$, $p_{61} = p_{16}$, $p_{23} = p_{32} = p_{34} = p_{43}$, etc.) means that the entries in the table above are sufficient to enable all the other coefficients to be obtained.

Table 7.5 DISTRIBUTION OF FREE BENDING MOMENTS

	Beam					
	1	2	3	4	5	6
Reaction ×Q	+1·246	+2·045	+0·805	-0·096	—	—
Free bending moment (kNm)	+1350	+2210	+870	-104	—	—
Distributed bending moment (kNm)						
+1350	+1030	+418	+54	-67	-67	-18
+2210	+685	+818	+596	+221	+22	-132
+870	+36	+235	+322	+235	+87	-45
-104	+5	-10	-28	-39	-28	-4
Totals	+1756	+1461	+944	+350	+14	-199

EXAMPLE 7.2   SIMPLY SUPPORTED BRIDGE SPANNING 36 m

General notes

Spans above about 25 m are probably most economically bridged by composite beams of which the steel component is an unsymmetrical welded plate girder having a smaller top than bottom flange. The economy of the girder can be improved by adopting a relatively thin web having a high depth-to-thickness ratio and by providing a web-stiffening system. Consideration might also be given to a steel box section composed of stiffened thin web and flange plates.

Selection of a trial beam is more difficult than for a symmetrical rolled beam section because of the increased number of variables involved and the lack of symmetry of the girder. A simple computer program is of use in calculating the properties of a number of trial sections and indeed an optimisation technique is available for minimum cost design.[14]

The remarks about transverse bracing and deck shuttering already made for short span bridges apply equally to this example. In addition, reduction of the bottom flange area over the ends (say quarter span) of the steel beam will produce further economy.

On spans of this length, HA loading is critical for girder design and so an HB load check is not carried out in the example.

Dimensions and material properties

Simple span	36·0 m
Beam spacing	2035 mm
Slab thickness	180 mm
28 day cube strength	$f_{cu} = 30$ N/mm^2
Mild steel	$f_y = 250$ N/mm^2

Loadings

Concrete	4·22 kN/m^2
Surfacing	2·94 kN/m^2

Design of girder

The design is based on a typical internal girder. Dead load and HA live load are transferred to each girder in proportion to the deck area supported. Three loading conditions are investigated.

Non-composite dead load
Slab $(4·22 \times 2·035) = 8·57$ kN/m $= 309$ kN
Beam $(2·70 \times 2·035) = 5·50$ kN/m $= 199$ kN

$\qquad\qquad\qquad$ Total $\quad$ 508 kN

Composite dead load

Surfacing $(2·94 \times 2·035) = 5·96 = 215$ kN

Live load
From HA loading data:
Uniformly distributed $= 7·50$ kN/m^2 (UD)

Knife edge   = 32·8 kN/m width

$UD = 7·50 \times 2·035$ = 15·4 KN/m   = 550 kN
$KE = 32·8 \times 2·035$ = 66·6 kN

Bending moments

                                                 (kN m)

$$M_{gs} = 0·125 \times 508 \times 36 = \qquad\qquad 2286$$
$$M_{gt} = 0·125 \times 215 \times 36 = \qquad\qquad 965$$
$$M_q = 0·125 \times 550 \times 36 = \quad 2470$$
$$+ \ 0·25 \times 66·6 \times 36 = \quad\ \ 600 \bigg\} \quad 3070$$

Shear

Shear values, both positive and negative, are required in order to compute the following:

(a) For proportioning the girder web and stiffening system, the *maximum* value (sum of composite and non composite)
(b) For calculating the number of shear connectors, the *range* of composite shear
(c) For checking the shear on possible shear planes, the maximum value of composite shear

While, in theory, it is necessary to draw shear force envelopes for the whole span, in practice, calculations of maximum values are only required for the ends, quarter points and centre of the girder.

From the shear force influence line (Figure 7.10),

$$v = q_x(x^2/2L) = + q_x x^2/2L$$
$$= -q_{(L-x)}[(L-x)^2/2L] = -q_{(L-x)}(L-x)^2/2L$$

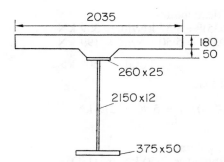

Figure 7.11 Midspan section of beam of Example 7.2. (Dimensions in mm)

where q_x and $q_{(L-x)}$ are the relevant rates of loading for loaded lengths x and $(L-x)$ respectively. The values of vertical shear calculated from the various loading conditions are shown in *Table 7.6*.

Steel beam dimensions

The beam shown in Figure 7.11 represents the result of possibly a number of trials. Certain constraints on the shape of the flanges in terms of their width-to-thickness ratios are imposed by codes of

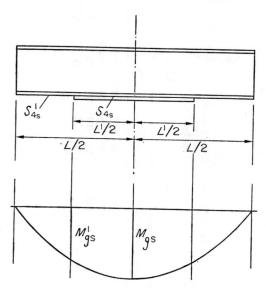

Figure 7.12 Reduction in bottom flange area

practice, and the web depth-to-thickness ratio will determine the necessity for stiffening the web plate. Cost studies appear to indicate that relatively thin and deep webs lead to economical beams even when account has been taken of the expense of providing a stiffener system.

Section properties of the steel section alone and of the composite beam for modular ratios of 7·5 and 15 are given in *Table 7.7*. From these the stresses at midspan have been calculated and recorded in *Table 7.8*. As is generally the case, the bottom flange tensile stress is critical. The working stress σ_4 is below the allowable value of 139 N/mm² but, as will be seen later, the difference may be required for HB loading.

Table 7.6 SHEAR FORCE VALUES

x Loaded length (m)	Dead load Steel +v (kN)	-v (kN)	Comp +v (kN)	-v (kN)	Rate (kN/m²)	Rate (kN/m)	Live load UD +v (kN)	-v (kN)	KE +v (kN)	-v (kN)	Max. on steel (kN)	Max. on composite DL (kN)	LL (kN)	Range on composite DL (kN)	LL (kN)
0		254		108			0	275	0	66·6	704	108	342	108	342
9		127		54	8·76	17·8	20	171	16·7	49·9	392	54	221	54	258
18	0		0		8·76	17·8	80	80	33·3	33·3	113	0	113	0	226
27	127		54		8·36	17·0	171	20	49·9	16·7	392	54	221	54	258
36	254		108		7·5	15·4	275	0	66·6	0	704	108	342	108	342

Table 7.7 PROPERTIES OF BEAM OF FIGURE 7.11

Steel beam

	1 A_s $\times 10^3$	2 y $\times 10^2$	3 $A_s y$ $\times 10^6$	4 y_s $\times 10^2$	5 I'_s $\times 10^9$	6 $A_s y_s^2$ $\times 10^9$	7 I_s $\times 10^9$
Top flange	6·5	2·43	1·58	13·6	—	12·0	12·0
Web	25·8	13·3	34·3	2·7	10·0	1·89	11·9
Bottom flange	18·8	24·3	45·8	8·3	—	13·0	13·0
Totals	51·1		81·7				36·9

$y_s = 81{\cdot}7 \times 10^6/51{\cdot}1 \times 10^3 = 1{\cdot}60 \times 10^3$

Composite beam

1 m	2 A_c $\times 10^5$	3 A_c/m $\times 10^4$	4 A_s $\times 10^4$	5 A_t $\times 10^4$	6 d_t $\times 10^3$	7 d_s $\times 10^2$	8 d_c $\times 10^2$	9 S_t $\times 10^7$	10 I_c/m $\times 10^8$	11 I_t $\times 10^{10}$
7·5	3·66	4·88	5·11	9·99	1·51	7·36	7·74	3·77	1·32	9·40
15·0	3·66	2·44	5·11	7·55	1·51	4·90	10·2	2·51	0·66	7·49

$d_t = 1600 - 90 = 1510$

$d_c = 5{\cdot}11 \times 1{\cdot}51 \times 10^7/9{\cdot}99 \times 10^4 = 7{\cdot}74 \times 10^2 \ (m = 7{\cdot}5)$

$\quad = 5{\cdot}11 \times 1{\cdot}51 \times 10^7/7{\cdot}55 \times 10^4 = 10{\cdot}2 \times 10^2 \ (m = 15)$

Section moduli

	1 y_1	2 y_2	3 y_3	4 y_4	5 S_1 $\times 10^7$	6 S_2 $\times 10^7$	7 S_3 $\times 10^7$	8 S_4 $\times 10^7$
Steel	1600	1420	1370	855	—	—	2·69	4·32
Composite $m = 7{\cdot}5$	864	684	634	1591	10·9	13·7	14·8	5·89
Composite $m = 15$	1110	930	880	1345	6·74	8·05	8·50	5·56

Table 7.8 STRESSES AT MIDSPAN

(N/mm²)

	Dead load		LL	Total
	Steel	Comp		
σ_1	—	0·955	3·76	4·72
σ_2	—	0·801	2·99	3·79
σ_3	80·5	11·3	20·8	112·6
σ_4	53·0	17·3	52·3	122·6

Reduction in bottom flange size

It is generally economical to make a reduction in the bottom flange area over about the first and last quarter of the span. An exact calculation of the theoretical length of reduced flange is tedious but the approximate method described below gives a fairly close estimate of the cut-off point for a given flange change.

It is assumed that the maximum bottom flange stress due to non-composite dead load at the point of maximum bending moment M_{gS} is equal to the associated stress at the point of cut-off, where the bending moment is M'_{gS}.

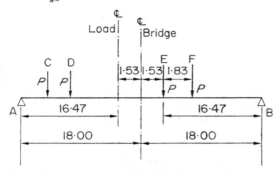

Figure 7.13 Longitudinal placing of HB *vehicle wheels. (Dimensions in* m*)*

Referring to Figure 7.12,

$$M'_{gS}/S'_{4s} = M_{gS}/S_{4s}$$

From the properties of the parabola

$$M'_{gS}/M_{gS} = 1-(L'/L)^2$$

Substituting for M'_{gS}/M_{gS}

$$L' = L(1-S'_{4s}/S_{4s})^{1/2}$$

Try a reduction in bottom flange size to 375×32 mm. From the section properties in *Table* 7.9,
  $L' = 36(1-3\cdot22/4\cdot32)^{1/2} = 18$ m (approx.), i.e. the reduction may take place at the quarter span points.

The stresses in the reduced section at the quarter span point are shown in *Table* 7.10 from which it will be seen that the initial assumption of equality of non-composite dead load stresses at level 4 is justified.

Table 7.9 PROPERTIES OF REDUCED SECTION

Steel beam

	1 A_s $\times 10^3$	2 y $\times 10^2$	3 $A_s y$ $\times 10^6$	4 y_s $\times 10^2$	5 I_t $\times 10^9$	6 $A_s y_s^2$ $\times 10^9$	7 I_s $\times 10^9$
Top flange	6·5	2·43	1·58	12·3	—	9·8	9·8
Web	25·8	13·3	34·3	1·4	10·0	0·51	10·5
Bottom flange	12·0	24·2	29·1	9·5	—	10·8	10·8
Totals	44·3		65·0				31·1

$y_s = 65/44\cdot3 = 1\cdot47 \times 10^3$

Composite beam

1 m	2 A_c $\times 10^5$	3 A_c/m $\times 10^4$	4 A_s $\times 10^4$	5 A_t $\times 10^4$	6 d_t $\times 10^3$	7 d_s $\times 10^2$	8 d_c $\times 10^2$	9 S_t $\times 10^7$	10 I_c/m $\times 10^8$	11 I_t $\times 10^{10}$
7·5	3·66	4·88	4·43	9·31	1·38	7·25	6·55	3·20	1·32	7·53
15·0	3·66	2·44	4·43	6·87	1·38	4·92	8·88	2·17	0·66	6·11

$d_t = 1470 - 90 = 1380$
$d_c = 4\cdot43 \times 1\cdot38 \times 10^7/9\cdot31 \times 10^4 = 6\cdot55 \times 10^2 \ (m = 7\cdot5)$
$\quad = 4\cdot43 \times 1\cdot38 \times 10^7/6\cdot87 \times 10^4 = 8\cdot88 \times 10^2 \ (m = 15)$

Section moduli

	1 y_1	2 y_2	3 y_3	4 y_4	5 S_1 $\times 10^7$	6 S_2 $\times 10^7$	7 S_3 $\times 10^7$	8 S_4 $\times 10^7$
Steel	1470	1290	1240	967	—	—	2·51	3·22
Composite $m = 7\cdot5$	745	565	515	1692	10·1	13·3	14·6	4·44
Composite $m = 15$	978	798	748	1459	6·25	7·67	8·16	4·19

Table 7.10 STRESSES IN REDUCED SECTION AT QUARTER SPAN

(N/mm²)

	Dead load		LL	Total
	Steel	Comp		
σ_1	—	0·774	3·11	3·88
σ_2	—	0·632	2·34	2·97
σ_3	68·2	8·87	28·6	105·7
σ_4	53·1	17·3	55·7	126·1

HB loading check

A typical internal beam will carry reactions from the HB vehicle of
P kN. Placing the vehicle longitudinally in the position to give the
worst bending moment (Figure 7.13) gives the reaction at B

$$R_B = 4P(16\cdot47/36\cdot00) = 1\cdot85\,P$$

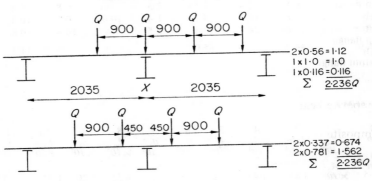

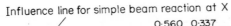

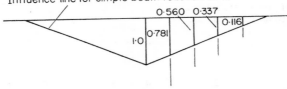

Figure 7.14 Transverse placing of **HB** *vehicle. (Dimensions in* mm)

and a maximum bending moment at E of

$$1\cdot85\,P\times16\cdot47 - P\times1\cdot83 = 28\cdot7\,P\ \text{kN m}$$

The transverse placing of the load to give the maximum value of P
is one of the two arrangements shown in Figure 7.13. For a wheel
load Q kN, assuming each piece of slab to be simply supported between longitudinal beams, it will be seen from Figure 7.14 that the
loading on beam X for either arrangement is

$$P = 2\cdot236\,Q$$

Because of the distribution properties of the deck this value may be
reduced. Using the reduction factor permitted by BS 153,

$$P = 0\cdot95\times2\cdot236\,Q = 2\cdot13\,Q$$

(a full distribution analysis could be carried out if desired)

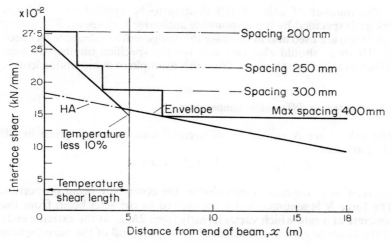

Figure 7.15. Interface shear envelope

The maximum bending moment in a typical internal girder thus becomes

$$M_{HB} = 2{\cdot}13 \times 28{\cdot}6\,Q = 61{\cdot}0\,Q \text{ kN m}$$

It is now necessary to check that the allowable stress in the bottom flange of the beam is not exceeded under HB loading.

Allowable bending stress in steel of yield stress 247 N/mm², thickness exceeding 19 mm, is 139 N/mm².

Dead load stresses at point E:

	(N/mm²)
Steel	53·0
Composite	17·3
Total	70·3

(Note. These stresses actually occur at *midspan* but the reduction at point E is small and has been ignored.)

An allowance of 25% increase in permissible bending stress is given for HB loading. Therefore the stress available for HB load is

$$1{\cdot}25 \times 139 - 70{\cdot}3 = 103{\cdot}6 \text{ N/mm}^2$$

from which

$$Q_{max} = 103{\cdot}6 \times 5{\cdot}89 \times 10^7 \times 10^{-6}/61{\cdot}0 \text{ kN}$$
$$= 100 \text{ kN}$$

The maximum HB vehicle class which this bridge can sustain is

$$45(100/112) = 40 \text{ units (imperial)}$$

The number of units of HB loading to be applied to a highway bridge is specified by the responsible authority. Commonly $37\,^1/_2$ units (imperial) are used and in this case the bridge will be adequate. A check of HB shear should also be made for the specified number of units although it has been omitted from this example to avoid undue length.

Shear caused by differential temperature

The axial force N caused by differential temperature is given in CP 117 part 2 as

$$N = A'\varepsilon(E_c I_c + E_s I_s)$$

where A' is a constant determined by the composite beam properties. The force N is assumed to be transferred to the steel beam from the concrete at a rate which varies linearly from $2N/L_s$ at the extreme ends of the beam to zero at a distance L_s from each end of the beam, where

$$L_s = 2\,[\mu A'(E_c I_c + E_s I_s)]^{1/2}$$
$$= 2\,(\mu)^{1/2}\,(N)^{1/2}/(E)^{1/2} = C_t(N)^{1/2}$$

For a given temperature range and shear connector constant, C_t is a constant. On the basis of a 10% temperature differential, and a coefficient of thermal expansion of $1 \cdot 1 \times 10^{-5}/\deg$ C the values of C_t are

Connector type	μ (mm^2/N)	C_t (mm N$^{1/2}$)
Stud	$2 \cdot 9 \times 10^{-3}$	$10 \cdot 3$
Other	$1 \cdot 45 \times 10^{-3}$	$7 \cdot 24$

The end shear rate $2N/L_s$ may be shown on the same basis to be

$$2N/L_s = 2N/C_t(N)^{1/2} = D_t(N)^{1/2}$$

Connector type	D_t (mm^{-1})
Stud	$0 \cdot 194$
Other	$0 \cdot 276$

Using these results for the particular beam

$$L_s = 10 \cdot 3 \times (2 \cdot 25 \times 10^5)^{1/2} = 4880 \text{ mm}$$

End shear $\quad = 0 \cdot 194 \times 4 \cdot 74 \times 10^2 = 93 \text{ N/mm} = 9 \cdot 3 \times 10^{-2} \text{ kN/mm}$

Shear connection

A full evaluation of the shear range to be transferred by shear connectors involves plotting a maximum shear range envelope for the effects of HA, HB and temperature shear. The HA shear force range has already been calculated. (*Table* 7.6). From it the interface shear values of *Table* 7.11 have been obtained. HB shear values have, for

Table 7.11 INTERFACE SHEAR DUE TO HA LOADING

x (m)	S_t/I_t (mm$^{-1}\times 10^{-4}$) $m = 7.5$	$m = 15$	Interface shear (kN/mm$\times 10^{-2}$) DL2	LL	Total
0	4·25	3·55	3·83	14·5	18·3
9	4·25	3·55	1·92	11·0	12·9
18	4·01	3·35	—	9·06	9·06

simplicity, not been calculated but it should be noted that for HB loading:

    (*a*) The allowable load on a shear connector is $0.40\times$ ultimate load and *not* $0.25\times$ ultimate load applicable to HA loading

    (*b*) The *maximum* shear value is used, not the shear range

The shear caused by differential temperature has already been calculated. Superimposing the HA and temperature shears and reducing the temperature shear by the 10% allowed in CP 117 leads to the shear envelope of Figure 7.15. Shear connectors and spacings can now be selected to transfer this shear.

Pairs of 19 mm diameter studs 102 mm long in concrete of strength 31 N/mm² have an ultimate capacity of 2×112 kN and a working capacity of $2\times112\times0.25 = 56$ kN. Their maximum spacing must not exceed

    (*a*) 600 mm
    (*b*) $3\times$ slab thickness $= 3\times180 = 540$ mm
    (*c*) $4\times$ connector height $= 4\times100 = 400$ mm (ruling value)

A possible arrangement of studs has been plotted on Figure 7.15 on the basis that

shear transfer capacity = working capacity/spacing

Spacing (mm)	Shear transfer capacity (kN/mm)
200	0·28
250	0·224
300	0·186
400	0·14

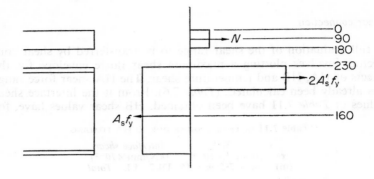

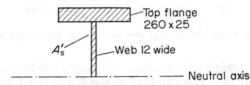

Figure 7.16 Ultimate resistance moment. (Dimensions in mm*)*

Temperature effects

(Calculated using the method of evaluation of CP 117 part 2 and the mean values of the section properties of the full and reduced beam sections shown in *Table* 7.12 with a modular ratio of 7.5)

Table 7.12 BEAM PROPERTIES FOR TEMPERATURE EFFECTS

d_t	I_o/m $\times 10^8$	I_s $\times 10^{10}$	A_o/m $\times 10^4$	A_s $\times 10^4$	$1/A'$ $\times 10^6$	A' $\times 10^{-7}$	A_c $\times 10^5$	S_{s3} $\times 10^7$	S_{s4} $\times 10^7$
1445	1·32	3·40	4·88	4·77	3·51	2·85	3·66	2·6	3·77

The free temperature strain due to a temperature differential of 10 deg C is $1\cdot1\times10^{-4}$. The sectional forces and moments calculated from the relationships in CP 117 are

$$N_c = N_s = N = 2\cdot25\times10^5 \text{ N}$$
$$M_c = 1\cdot26\times10^6 \text{ N mm}$$
$$M_s = 3\cdot24\times10^8 \text{ N mm}$$

from which the stress changes due to the temperature differential may be calculated. The sense of N_c, N_s, M_c and M_s will depend on whether the beam is warmer than the slab or vice versa.

The absolute values of the stresses are given in *Table* 7.13 and the

Table 7.13 ABSOLUTE VALUES OF TEMPERATURE STRESSES

(N/mm²)

N/A_c	$2 \cdot 25 \times 10^5 / 3 \cdot 66 \times 10^5 =$	$0 \cdot 615$
N/A_s	$2 \cdot 25 \times 10^5 / 4 \cdot 77 \times 10^4 =$	$4 \cdot 72$
M_c/S_{c1}	$1 \cdot 26 \times 10^6 / 1 \cdot 1 \times 10^7 =$	$0 \cdot 114$
M_s/S_{s3}	$3 \cdot 24 \times 10^8 / 2 \cdot 6 \times 10^7 =$	$12 \cdot 5$
M_s/S_{s4}	$3 \cdot 24 \times 10^8 / 3 \cdot 77 \times 10^7 =$	$8 \cdot 6$

resultant stresses at the four levels in the composite beam for the two cases (slab warmer or beam warmer) in *Table* 7.14. In either case it

Table 7.14 TEMPERATURE STRESSES AT LEVELS 1 TO 4 FOR CASE OF SLAB OR BEAM WARMER

(N/mm²)

Stress	Slab warmer			Beam Warmer		
	N/A	M/Z	Total	N/A	M/Z	Total
σ_1	$-0 \cdot 615$	$+ 0 \cdot 114$	$- 0 \cdot 501$	$+0 \cdot 615$	$- 0 \cdot 114$	$+ 0 \cdot 501$
σ_2	$-0 \cdot 615$	$- 0 \cdot 114$	$- 0 \cdot 729$	$+0 \cdot 615$	$+ 0 \cdot 114$	$+ 0 \cdot 729$
σ_3	$+4 \cdot 72$	$+12 \cdot 5$	$+17 \cdot 2$	$-4 \cdot 72$	$-12 \cdot 5$	$-17 \cdot 2$
σ_4	$+4 \cdot 72$	$- 8 \cdot 6$	$- 3 \cdot 88$	$-4 \cdot 72$	$+ 8 \cdot 6$	$+ 3 \cdot 88$

will be seen that the stress in the bottom flange of the steel beam (normally critical) is very little affected by temperature differential.

Ultimate load check

Although ultimate load design methods are not permitted for bridges it is useful to check the load factor available against collapse.

Assuming that the neutral axis lies in the steel beam (Figure 7.16):

Force in slab $N(= 3 \cdot 66 \times 10^5 \times 0 \cdot 444 \times 30) = 4 \cdot 87 \times 10^6$ N

Force in beam $A_s f_y(= 5 \cdot 11 \times 10^4 \times 250)$ $\qquad = 1 \cdot 28 \times 10^7$ N

$2A' f_y(= (12 \cdot 8 - 4 \cdot 87) \times 10^6)$ $\qquad = 7 \cdot 9 \times 10^6$ N

$A'(= 7 \cdot 9 \times 10^6 / 2 \times 250)$ $\qquad = 1 \cdot 58 \times 10^4$ mm²

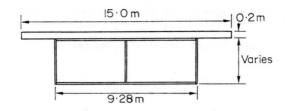

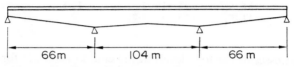

Figure 7.17 Continuous composite box beam–Example 7.3

showing that the assumption of the neutral axis position is correct.

From the geometry of the steel section:
Area of top flange ($= 260 \times 25$) $= 0{\cdot}65 \times 10^4$ mm²
Neutral axis lies in web at a point

$(1{\cdot}58 - 0{\cdot}65) \times 10^4/12 = 775$ mm below top flange underside

Centroid of A' lies

$((0{\cdot}65 \times 10^4 \times 12{\cdot}5 + 0{\cdot}93 \times 10^4 \times 413)/1{\cdot}58 \times 10^4)$

$= 241$ mm below top of top flange

Taking moments about the line of action of N, the ultimate resistance moment of the composite section is:

$1{\cdot}28 \times 10^7 \times 1510 = 1{\cdot}93 \times 10^{10}$
$-7{\cdot}9 \times 10^6 \times 381 \quad = 0{\cdot}3 \times 10^{10}$
$\qquad$ Total $\qquad \overline{1{\cdot}63 \times 10^{10}}$ N mm $= 16\ 300$ kN m

$1{\cdot}5 \times$ dead load bending moment $= 1{\cdot}5 \times 3245 = 4870$
$2{\cdot}0 \times$ live load bending moment $\ = 2{\cdot}0 \times 3070 = 6140$
$\qquad\qquad\qquad\qquad$ Total $\quad \underline{11\ 010}$ kN m

As an ultimate load basis the beam has an adequate load factor against collapse.

EXAMPLE 7.3   THREE SPAN CONTINUOUS BRIDGE 66–104–66 m

General notes

For spans of this length it is not possible to say that any particular structural solution is the correct one. There is, for example, an immediate choice between a continuous beam or a cantilever and suspended span. The solution given here therefore should be treated rather as an illustration of general methods of calculation of a long span composite beam than as necessarily the best solution for a crossing of such magnitude.

Again the choice of a box beam is to some extent arbitrary, though their use has become common in recent years. No attempt is made here to deal with the many special design aspects of steel box girders, notably the analysis of twisting due to unsymmetrical loading and the provision of suitably stiffened plates in webs, flanges and diaphragms. These matters have received much attention and reference should be made to the many technical papers and reports on the subjects.

The complications of negative (hogging) bending in continuous composite beams may be avoided by the adoption of a method by which sufficient precompression is introduced into the slab to ensure that the net stress in it does not exceed the (small) allowable tensile stress under any condition of loading. In this scheme, cable prestressing in the deck is used. The best advantage from cable prestressing is obtained when the deck is not connected to the steel beam until prestressing is completed. However, this leads to certain complications if the deck slab is not to buckle under the prestressing force, and so the less efficient method of prestressing after composite action has been adopted.

It must be borne in mind that the loading rates in BS 153 part 3*a* are strictly applicable to *simply supported* bridges of span up to about 100 m. They can be used for continuous bridges with the proviso that care must be taken to make whatever amendments are necessary for fixity at the supports, continuity or other indeterminate or special conditions. The definition of 'loaded length' is important in continuous beams, in which the bending moment influence line for a particular point may consist of parts of different sign having unequal length. The maximum bending moment will then be found by consideration of any part or combination of separated parts using the loading appropriate to the length or the total combined length of the loaded portion.

For spans of this length, normal HA loading is likely to be critical, at least for overall bending and shear. For this reason HB loading has not been considered: it should be remembered, however, that the local effects of HB loading may well be more severe than those of HA.

Calculation of bending moments

The method of influence coefficients is well adapted to the evaluation of bending moments in continuous beams of variable section; numerical integration is performed using Simpson's rule. Bending moments are calculated for the following stages in the construction sequence:

1 Erect steel box girders
2 Concrete deck slab over interior supports
3 Prestress slab
4 Concrete remainder of deck slab
5 Lay surfacing
6 Apply live load

The notation adopted for calculation of influence coefficients is that of Morice.[15] To avoid too great a concentration of numerical working, two loading cases only are evaluated in full; the remainder are shown diagrammatically.

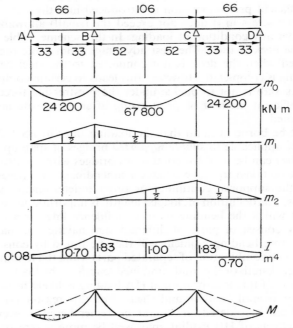

Figure 7.18 Stage 1 bending moment–self-weight of steel box beam. (Dimensions in m)

Dimensions and material properties

Three continuous spans 66–104–66 m
Twin cell steel box girder of total width 9·28 m supporting concrete
slab 200 mm thick. Box depth varies (Figure 7.17).
High yield steel $f_y = 350$ N/mm²

Loadings

Steel box girder weight varies
Concrete 70·8 kN/m run of box beam
Live load HA to British Standard 153

Loading case 1. Steel beam erected

The free bending moment diagrams caused by the self-weight of
the steel box are shown in Figure 7.18 with moment releases at B and
C.

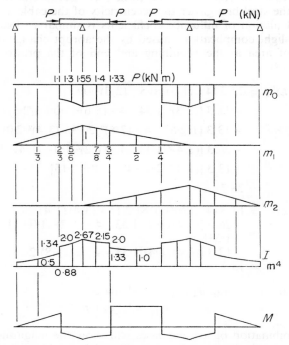

Figure 7.19 Stage 3 *bending moment–prestress slab. (Dimensions in* m*)*

$$Ef_{11} = Ef_{22} = 33/3 \, (0+4\times0{\cdot}5\times0{\cdot}5/0{\cdot}7+1\times1/1{\cdot}83)$$
$$+52/3 \, (1\times1/1{\cdot}83+4\times0{\cdot}5\times05{\cdot}/1{\cdot}0+0)$$
$$= 48{\cdot}4$$
$$Ef_{21} = Ef_{12} = 52/3 \, (4\times0{\cdot}5\times0{\cdot}5/1{\cdot}0) = 17{\cdot}4$$
$$Eu_1 = Eu_2 \;\; = -33/3 \, (4\times0{\cdot}5\times24\,200)-52/3 \, (4\times0{\cdot}5\times67\,800)$$
$$= 3{\cdot}23\times10^6$$

Because of symmetry $x_1 = x_2$ (bending moments at B and C)

$$x_1 = x_2 = -3{\cdot}23\times10^6/(48{\cdot}4+17{\cdot}4) = 4{\cdot}92\times10^4 \text{ kN m}$$

Loading case 3.   Prestress slab over supports

The amount of prestressing force required to eliminate tension in the deck (or to keep tensile stresses below a specified level) is determined from consideration of all loading conditions. In this case it has been decided to prestress 48 m of slab over each support: 22 m in the end spans and 26 m in the centre span. A straight cable located at the centroid of the concrete slab is to be used. Because of the varying depth of the steel box girder the eccentricity of the cable also varies.

The calculation of influence coefficients is similar to loading case 1 with a slight complication caused by the abrupt change of second moment of area at the beginning and end of the prestressed slab (Figure 7.19).

$$Ef_{11} = Ef_{22} = 22/3 \, [4\times(1/3)^2/0{\cdot}5+(2/3)^2/0{\cdot}88]$$
$$+11/3 \, [(2/3)^2/1{\cdot}34+4\times(5/6)^2/2{\cdot}0+1/2{\cdot}67]$$
$$+13/3 \, [1/2{\cdot}67+4\times(7/8)^2/2{\cdot}15+(3/4)^2/2{\cdot}0]$$
$$+26/3 \, [(3/4)^2/1{\cdot}33+4\times(1/2)^2/1{\cdot}0+(1/4)^2/1{\cdot}33]$$
$$+13/3 \, [(1/4)^2/2{\cdot}0+4\times(1/8)^2/2{\cdot}15]$$
$$= 38{\cdot}7$$
$$Ef_{12} = Ef_{21} = 13/3 \, [4\times(7/8)\,(1/8)/2{\cdot}15+(3/4)\,(1/4)/2{\cdot}0]$$
$$+26/3 \, [(3/4)\,(1/4)/1{\cdot}33+4\times(1/2)^2/1{\cdot}0+(1/4)\,(3/4)/$$
$$1{\cdot}33]$$
$$+13/3 \, [(3/4)\,(1/4)/1{\cdot}33+4(7/8)\,(1/8)/2{\cdot}15]$$
$$= 13{\cdot}2$$

In a similar way u_1 and u_2 may be calculated

$$Eu_1 = Eu_2 = -30{\cdot}1 \, P$$

The combination of loading cases will show the magnitude of the maximum tensile stress in the concrete which would exist in the ab-

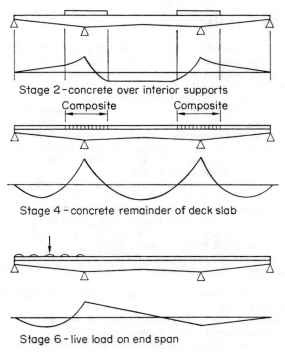

Figure 7.20 Bending moment diagrams for loading stages 2,4 and 6

sence of prestressing. The value of P can then be adjusted to eliminate the tension.

Loading cases 2, 4 and 6 are shown diagrammatically in Figure 7.20.

Shear connection, creep and shrinkage, temperature stresses and other aspects of the design of a long span bridge are determined by the general criteria already described in Examples 7.1 and 7.2, complicated necessarily by the continuous nature of the beam.

194   COMPOSITE BRIDGES

REFERENCES   CHAPTER 7

1. ELLIOTT, P., 'Can steel bridges become more competitive?', *Proc. British Constructional Steelwork Association Conf. on Steel Bridges* (June 1968)
2. SIMPSON, C. V., 'Modern long span steel bridge construction in Europe', *Proc. Instn. civ. Engrs, Suppl.* (ii) (1970)
3. BECKETT, D., *Bridges*, Hamlyn, London (1970)
4. 'Bridges in composite construction', British Constructional Steelwork Association, London (1967)
5. HAWRANEK, A., and STEINHARDT, O., *Theorie und berechnung der stahlbrucken* (Theory and calculation of steel bridges), Springer-Verlag, Berlin (1958)
6. 'Criteria for the assessment of box girder bridges — interim rules', Department of the Environment., London (May 1971).
7. British Standard 153, '*Specification for steel girder Bridges, Part 3a, Loads*, British Standards Institution, London (1966)
8. SAWKO, F., 'Bridge deck analysis — electronic computers versus distribution methods', *Civ. Engng publ. Wks Rev.*, London (Apr. 1965)
9. HENDRY, A. W., and JAEGER, L. G., *The analysis of grid frameworks and related structures*, Chatto and Windus, London (1958)
10. British Standard Code of Practice CP 117: Part 2, *Composite construction in structural steel and concrete: Beams for bridges*, British Standards Institution, London (1967)
11. FLINT, A. R., and EDWARDS, L. S., 'Limit state design of highway bridges', *Struct. Engr* **48**, No. 3, (Mar. 1970)
12. *Simply supported bridges in composite construction*, British Constructional Steelwork Association, London, Publication BD2 (1970)
13. PUCHER, A., *Influence surfaces of elastic plates*, Springer-Verlag, Berlin (1964)
14. REGE, P.M.,' Optimisation of composite steel and concrete construction', MSc Thesis, University of Surrey, Guildford (1970)
15. MORICE, P. B., *Linear structural analysis*, Thames and Hudson, London (1962)

Notations

Systems of notation for composite construction vary not only internation-
ally but even between the two parts of the current (1972) British Standard.
The symbols listed here have been based as far as possible on the notations
prepared by the Comité Européen du Béton presented in *Standard Notations* by B. H. Spratt (Cement and Concrete Association, Tech. Rep.
42.459, London, 1971). There are, of necessity, occasions when this notation cannot be followed; such cases are met by a definition in the text at
the appropriate place.

Upper and lower case Italic letters

A	cross-sectional area
b	width
d	depth
E	modulus of elasticity
e	eccentricity
f	strength
g	distributed dead load
h	depth of composite neutral axis below top of slab
I	second moment of area
k	coefficient
L	span
M	bending moment
m	modular ratio
N	axial force
P	prestressing force
Q	live load
q	distributed live load
R	reactive force
r	radius
S	first moment of area
s	slip

V shear force
v shear force per unit length or width

General subscripts

c   concrete, compression
e   elastic, effective
g   dead load
i   initial
p   prestress
q   live load
s   steel
t   composite, tension
u   ultimate
v   shear

Lower case Greek letters

α coefficient
β coefficient
δ deflection
ε strain
$\varkappa$ curvature
λ coefficient

μ shear-connector modulus
ϱ relaxation coefficient
σ normal stress
τ shear stress
ϕ creep coefficient
ψ coefficient

Index

Floors, filler joist, 4
fireproof, 4
Friction-grip bolts, high-strength, 103–104, 112

Girders, delta top, 120
lattice. *See* Composite lattice girders
open web, 121
plate. *See* Composite plate girders
Gluing for shear connection, 104

Haunch dimensions, 105
Haunched sections, 133
High-strength friction-grip bolts, 103–104, 112
Hogging, 51–53, 82, 189
Hognestad parabolic stress–strain curve, 125

Influence coefficients, 190
Instantaneous modulus of elasticity, 20

Jacking, prestressing by, 73–78, 88–91, 137

Kane system, 119

Lattice girders. *See* Composite lattice girders
Load–slip characteristic, 15
Longitudinal shear failure, 105
Longitudinal shear strength, mean ultimate, 105

Materials, properties of, 11
Modified Goodman diagram, 97
Modular ratio, 33, 71

Negative bending, 82–91, 189
Non-composite action, 82
Notation, 195

Open web girders, 121

Partial interaction, 26
Phased concreting, 71
Plastic design method, 133

Plastic hinges, 51
Plate girders. *See* Composite plate girders
Plates, composite, 129–130
Pont de la Madeleine, 158
Prefabrication, 132
Preflex beams, 127–128, 135
Prestressed composite tension flanges, 119
Prestressing, 9, 72, 135
by cables, 78–81, 83–86
by cambering, 88–91
by jacking, 73–78, 88–91, 137
examples, 75, 79, 86
with temporary link system, 89
Prestressing steel, 13
Properties of materials, 11
Propped construction, 7–8, 60, 63, 69, 132, 138, 146
Pull-out test, 93
Push-out test, 93, 98

Raith Bridge, 156
Reduction factor, 50
Reinforced concrete, early development, 2
Reinforcement, additional, 83
transverse, 105–106, 151
Relaxation, 20, 24
Relaxation characteristics, 42
Relaxation coefficient, 34–35
Relaxation method, 34, 42, 48
Residual stresses, 75
Restoring moment, 49
Ribbed sheet steel shuttering, 133–135

Sagging, 51–53
Sattler, approximate method, 33, 41, 48
approximate relationships, 34
Section modulus, 33, 74
Section modulus ratio, 75
Sectional forces, 31
Shear, due to differential strain, 107, 117
due to differential temperature, 184
and shrinkage, 115
in bridge design, 176
Shear connection, 4, 15, 92–118
deformation of, 26
in bridges, 185
in buildings, 149

200 INDEX